DEFORMATION AND FRACTURE OF SOLIDS

DEFORMATION AND FRACTURE OF SOLIDS

ROBERT M. CADDELL

Professor of Mechanical Engineering
The University of Michigan

PRENTICE-HALL, INC.

Englewood Cliffs, New Jersey 07632

Library of Congress Cataloging in Publication Data

CADDELL, ROBERT M
 Deformation and fracture of solids.

 Bibliography: p.
 Includes index.
 1. Deformations (Mechanics) 2. Fracture
mechanics. I. Title.
TA417.6.C32 1980   620.1'123   79–20255
ISBN 0-13-198309-1

$$T A$$
$$4 17. 6$$
$$C 3 2$$

Printed in the United States of America
10 9 8 7 6 5 4 3 2 1

Editorial Production Supervision and interior design by Theodore Pastrick
Interior artwork by Reproduction Drawings, Ltd.
Manufacturing Buyer: Gordon Osbourne

PRENTICE-HALL INTERNATIONAL, INC., *London*
PRENTICE-HALL OF AUSTRALIA PTY. LIMITED, *Sydney*
PRENTICE-HALL OF CANADA, LTD., *Toronto*
PRENTICE-HALL OF INDIA PRIVATE LIMITED, *New Delhi*
PRENTICE-HALL OF JAPAN, INC., *Tokyo*
PRENTICE-HALL OF SOUTHEAST ASIA PTE. LTD., *Singapore*
WHITEHALL BOOKS LIMITED, *Wellington, New Zealand*

4-9-80 2x

CONTENTS

PREFACE   ix

CONVERSION FACTORS AND SI UNITS   xii

1. STRESS   1

    1–1   Introduction   *1*
    1–2   A Generalized Three Dimensional Analysis of
           a Homogenous Stress State   *5*
    1–3   Stress Transformations   *11*
    1–4   Mohr's Circle—Its Use and Limitations   *15*
           PROBLEMS   *28*

2. STRAIN   32

    2–1   Introduction   *32*
    2–2   The Two Dimensional Case for Small Strains   *33*
    2–3   The Three Dimensional Case for Small Strains   *36*

2–4   Strain Transformations   *38*
2–5   Mohr's Circle for Strains   *40*
2–6   Engineering Strain and True Strain   *42*
      REFERENCES   45
      PROBLEMS   45

3. *ISOTROPIC ELASTICITY*   **47**

3–1   Introduction   *47*
3–2   Model of a Perfectly Elastic Solid   *48*
3–3   Constitutive Relations   *49*
3–4   Elastic Constants   *50*
3–5   Relationship of Mohr's Circle for Stress and Strain   *52*
3–6   Strain Energy Due to the Work of Elastic Deformation   *55*
3–7   Two Important Physical Situations—Plane Stress
      and Plane Strain   *56*
      REFERENCES   56
      PROBLEMS   56

4. *MACROSCOPIC PLASTICITY*   **58**

4–1   Introduction   *58*
4–2   Comparison of Elasticity and Plasticity   *59*
4–3   Models for Plastic Deformation   *60*
4–4   The Yield Locus and Surface   *63*
4–5   Yield Criteria   *66*
4–6   Tresca Criterion   *67*
4–7   von Mises Criterion   *69*
4–8   Distortion Energy   *74*
4–9   Octahedral Shear Stress   *75*
4–10 Flow Rules or Plastic Stress-Strain Relationships   *76*
4–11 Normality and Yield Surfaces   *81*
4–12 Plastic Work   *86*
4–13 Comparison of Mohr's Circles for Stress and
      Plastic Strain Increments   *87*
      REFERENCES   88
      PROBLEMS   88

5. *OBSERVED MACROSCOPIC BEHAVIOR OF
    DUCTILE METALS*   **92**

5–1   Introduction   *92*
5–2   Deformation of Ductile Metals by Uniaxial Tension   *93*
5–3   Load versus Elongation   *93*
5–4   Engineering or Nominal Stress and Strain   *96*
5–5   Measurement of Yield Stress   *98*
5–6   Indications of Ductility   *99*
5–7   True Stress and True Strain   *101*

5–8   Work Hardening by Uniaxial Tension   *102*
5–9   Determination of the Work-Hardening Equation   *105*
5–10   Behavior After Necking   *110*
5–11   Balanced Biaxial Tension (Bulge Test)   *114*
5–12   Plane Strain Compression   *115*
5–13   Ductility of Prior Worked Metals   *117*
5–14   Summary   *120*
       REFERENCES   *120*
       PROBLEMS   *121*

6. *VISCOELASTICITY*   **124**

6–1   Introduction   *124*
6–2   Simple Model Depicting Time-Dependent or
       Viscous Behavior   *127*
6–3   Compound Models   *128*
6–4   Maxwell Model   *128*
6–5   Voigt or Kelvin Model   *132*
6–6   Three-component Model   *137*
6–7   Four-component Model   *139*
       PROBLEMS   *140*

7. *ELEMENTS OF DISLOCATION THEORY*   **143**

7–1   Introduction   *143*
7–2   Maximum Theoretical Shear Stress   *144*
7–3   Models of Pure Edge and Pure Screw Dislocations   *147*
7–4   Dislocation Motion and the Burgers Vector   *148*
7–5   Mathematical Derivation of Stresses and Strains Caused
       by Dislocations   *152*
7–6   Shear Strain Caused by Dislocation Movement   *160*
7–7   Strain Energy Caused by Dislocations   *162*
7–8   Force Acting on a Dislocation Due to Externally
       Applied Stresses   *164*
7–9   Forces Due to Interaction of Dislocations   *166*
7–10   Generation of Dislocations   *172*
7–11   Climb and Cross Slip   *175*
7–12   Qualitative Explanations of Macroscopic Behavior   *177*
       REFERENCES   *184*
       PROBLEMS   *188*

8. *FRACTURE AND FRACTURE MECHANICS*   **193**

8–1   Introduction   *193*
8–2   Modes or Types of Fracture   *194*
8–3   Maximum Theoretical Cohesive Strength of Solids   *197*
8–4   Strain Energy Release Rate   *205*
8–5   Design Considerations   *207*

8–6   Linear Elastic Fracture Mechanics   *208*
8–7   Tie-in Between Strain Energy Release Rate and
        the Stress Intensity Factor   *223*
8–8   Fracture Toughness—The Gurney Approach   *224*
8–9   Crack Stability in Physical Terms   *233*
8–10   Crack Stability in Mathematical Terms   *235*
8–11   Initial Crack Velocities   *238*
8–12   Influence of Specimen Size on Fracture Toughness
        Measurements   *239*
        REFERENCES   244
        PROBLEMS   245

9. *COMPOSITES*   **252**

9–1   Introduction   *252*
9–2   Definition of Composites   *253*
9–3   Continuous Fiber Composites and the Rule of
        Mixtures   *254*
9–4   The Modified Rule of Mixtures (MROM)   *257*
9–5   Discontinuous Fiber Composites   *258*
9–6   Concept of the Average Fiber Stress   *262*
9–7   General Comments   *264*
        REFERENCES   269
        PROBLEMS   270

10. *FATIGUE*   **272**

10–1   Introduction   *272*
10–2   Factors that Influence Fatigue   *273*
10–3   Macroscopic Design   *279*
10–4   Fatigue and Fracture Mechanics   *289*
10–5   Crack Propagation   *289*
10–6   Failure Analysis   *294*
        REFERENCES   299
        PROBLEMS   300

INDEX   **303**

PREFACE

The mechanical behavior of solids, as expressed by deformation and fracture, is a field of fundamental importance in engineering. Yet, most undergraduate curricula provide only a cursory exposure to this subject. This conclusion stems from my personal experience that includes over twenty years of teaching both undergraduate and graduate students from a variety of engineering disciplines and whose prior background was obtained at numerous universities and colleges. My major teaching involvement with this subject has been with an elective course that is populated with seniors and first-year graduate students. They have studied in such fields of engineering as Mechanical, Naval Architecture, Materials and Metallurgy, Aerospace and Applied Mechanics. Perhaps the major challenge that arises in such a situation is to accommodate the variety of backgrounds of such a diverse group of individuals. Most have had a course in the Science of Engineering Materials and a traditional Strength of Materials approach to design. Some have pursued further work in metallurgy, mechanical design, and stress analysis; however, practically none has been exposed to the important subject of fracture. Yet even in regard to the most basic concepts that have been covered in earlier studies, I have found an unawareness and confusion that is perplexing but which must be overcome before proceeding with other topics. The format and content of this book constitute an approach that has been successful in practice.

Chapters 1 through 3 provide a concise review of stress, strain and elasticity that should clarify any misconceptions that students usually exhibit. The use of Mohr's circle, is more detailed than one would find in most similar texts, the reason arising from the observation that most students neither appreciate the limitations of this technique nor fully understand the basis behind it. Because traditional courses in design and materials have stressed elastic behavior of solids, the importance of plasticity is usually ignored. For that reason, Chaps. 4 and 5 contain more extensive coverage than is usually found in other texts of this nature. Time-dependent behavior is discussed in Chap. 6 where the main emphasis is on the importance of rate equations in the study of all the parameters that are influenced by time effects. Chapter 7 is not intended to satisfy experts in dislocation theory but the approach taken does provide a useful and direct way to explain many important macroscopic observations of importance to engineers. The major concepts of fracture and fracture mechanics are covered in Chap. 8. This has become such an important subject in certain areas of modern design that it should be a part of undergraduate study, yet it receives little if any attention at most schools. Besides covering the concepts and use of the stress intensity factor and strain energy release rate, an approach using an energy balance is presented. The latter is not found in other books yet it provides a better introduction to the physical meaning of fracture toughness as determined by experiment. Chapter 9 introduces some elementary ideas in regard to composite materials. There is no doubt that their use in engineering applications will increase in the years ahead and the brief coverage in this chapter will point out their desirable features as well as limitations. A broad overview of fatigue constitutes Chap. 10. Both the traditional design approach and the more recent concepts involving fracture mechanics are covered to a meaningful extent. A brief summary of failure analysis completes this chapter.

Having used several other texts in the past, the opinions and complaints of many students have guided me in writing this book. Often, because of their length, similar texts cannot be covered reasonably in a single semester. Yet students must pay for the full book, usually with some chagrin. To overcome this complaint, I have attempted to pare topics to a minimal but essential level and have omitted some completely. This is more of a personal choice than an implication that such topics are of lesser importance. Any teacher using this book can easily fill this void with a minimal number of handout notes. In response to other student complaints, a number of example problems are provided in each chapter and, where necessary, derivations are included in full measure. Numerous end-of-chapter problems include some that are directed towards basic definitions and others that pose more of a challenge.

Both the English and SI unit notations are used and important conversion factors are summarized after this preface. If one is objective, there is no question that the SI system is more convenient. To a large extent a bias for English units, based upon nothing more than tradition, must be overcome. Interspersing the use of both systems should assist in reducing reluctance towards what appears to be inevitable.

Most problems are self-contained with regard to property values needed for a solution, but in a number of instances deliberate omissions have been made. This forces students to check other sources and anyone who has practiced engineering will agree that this is often a critical part of real problem solving. Students must be made aware of this and the sooner the better.

References are limited to papers on basic developments or to other texts that provide further detail on topics of concern. Except in research problems, students seem disinterested in checking references. By limiting the number, it is hoped that these younger readers will develop this important habit.

My indebtedness to certain colleagues and former students must now be paid. In particular Drs. A. G. Atkins, D. K. Felbeck, W. F. Hosford Jr., K. C. Ludema, and Y. W. Mai have provided many incisive comments and discussions during our years of acquaintance. To segregate their inputs in a specific way is an impossible task; it must suffice to pay my thanks to each of them. Of the many students who have given constructive aid I must mention Drs. R. S. Raghava and A. R. Woodliff. They have provided the type of student input that all teachers desire. Finally I must thank Ms. Karen Almas, Ms. Mary Anne Brocious, Ms. Gloria Hartman, and Mrs. Karen Chapin for the many hours of excellent assistance they provided in the typing of this text. Their patience and care have been invaluable.

Robert M. Caddell
Ann Arbor, Michigan

Some Basic Units and Their Abbreviations as Specified for The International System of Units (SI)

Unit	Standard	Abbreviation
length	meter	m
mass	kilogram	kg
time	second	s
*force	Newton	$N = kg \cdot m/s^2$
*stress	Newton/meter^2	N/m^2
*stress or pressure	Pascal $= 1\ N/m^2$	Pa
*energy	Joule	$N \cdot m$

These are derived from basic units.

Multiplication Factors Used in the SI System

Factor	Prefix	Symbol	Factor	Prefix	Symbol
10^{12}	tera	T	10^{-6}	micro	μ
10^{9}	giga	G	10^{-9}	nano	n
10^{6}	mega	M	10^{-12}	pico	p
10^{3}	kilo	k	10^{-15}	femto	f
10^{-3}	milli	m	10^{-18}	atto	a

Useful Conversion Factors—English to SI Units

To convert from	to	multiply by
inch	meter	2.54×10^{-2}
feet	meter	3.048×10^{-1}
inch^2	meter^2	6.452×10^{-4}
feet^2	meter^2	9.29×10^{-2}
inch^3	meter^3	1.639×10^{-5}
feet^3	meter^3	2.832×10^{-2}
pound-force	Newton	4.448
pounds/inch^2	Newton/meter^2	6.895×10^{3}

Other Useful Relations

1 micron $= 10^{-4}$ cm $= 10^{-6}$ m
1 Angstrom $= 10^{-8}$ cm $= 10^{-10}$ m
1 dyne/cm^2 $= 1.44 \times 10^{-5}$ pounds/inch^2
10 Angstroms $= 1$ nm
For body-centered cubic unit cell, $a_0 = 4r/\sqrt{3}$
For face-centered cubic unit cell, $a_0 = 4r/\sqrt{2}$
  where $r =$ atomic radius and $a_0 =$ lattice parameter

1

STRESS

1-1 INTRODUCTION

Stress is defined as the intensity of force per unit area and many problems
have been solved with the use of this simple concept. Experiments involving
uniaxial tension, compression and torsion are most often used to provide a
physical illustration of the above definition. Yet this introductory concept
does not provide a full appreciation of the nature of stress. Biaxial or triaxial
states of stress are often encountered and excessive emphasis on one-dimen-
sional situations can lead to decided misconceptions.

Initially we used the words *force* and *area* above; both are vector quantities
requiring magnitude and direction for their full description. Stress requires
two sets of direction cosines for its complete description and if it is desired
to transform the stress components from one set of coordinate axes to a different set, then certain mathematical *rules* must be followed. This is called
tensor transformation and a knowledge of these rules and accompanying

notation provides great convenience, since simple shorthand expressions are used to describe the state of stress. A background in tensor analysis is not essential for the purposes of this text but it is important to realize that:

1. *Stress* is a second-order tensor (usually just called a tensor) involving two sets of direction cosines for transformation purposes.
2. *Force* is a first-order tensor (usually called a vector) involving one set of direction cosines for transformation purposes.
3. *Temperature* is a zero-order tensor (usually called a scalar) and is independent of any direction notation.

Other physical examples could be used, but the major purpose of these remarks is to indicate that stress is *not* a vector.

Situations can be visualized where the state of stress varies throughout a body subjected to loading. Consider the simple tension test of a ductile metal after necking has occurred. Although the *force* is common throughout the length, the change in area (plus the geometry) causes the stress to vary from point to point. To accommodate this variation the concept of the *state of stress at a point* is used.

Consider an elemental force, *dF*, acting at a point *P* included in an elemental area, *dA*, as shown in Fig. 1-1(a). In the most general sense, *dF* is neither

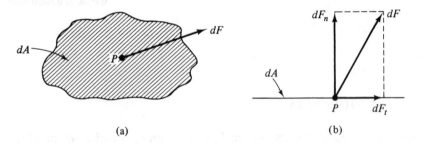

(a)   (b)

Figure 1-1 Forces acting on an elemental area showing total force (a) and resolved forces (b).

normal to *dA* nor parallel to it, as in Fig. 1-1(b), and *dF* may be reduced to the normal and tangential components dF_n and dF_t. As $dA \rightarrow 0$ in the limit, stress is *defined* as follows:

$$\sigma = \frac{dF}{dA}, \qquad \sigma_n = \frac{dF_n}{dA}, \qquad \sigma_t = \frac{dF_t}{dA}$$

where σ is the *total state of stress* at *P* and σ_n and σ_t are the normal and tangential components. Note that the same *area* is involved in each of the above definitions.

Now if P is envisioned to lie at the centroid of a very small parallelepiped (this is a mathematical concept only) where any changes across this model are very small, the state of stress at point P is considered to be homogeneous and is described by the model. If numerous forces act on a body, their effect at P (using an arbitrary coordinate system) is described by Fig. 1-2. For

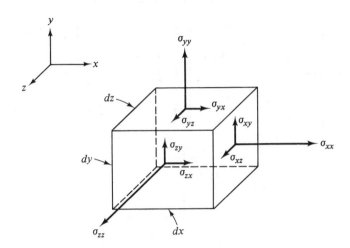

Figure 1-2 Stress element for a homogeneous state of stress at a point. As shown, all stress components are positive by convention.

stationary equilibrium, each face opposite to the three shown has identical stresses acting such that a force balance would give $\sum F_x = 0$ while the absence of rotation means $\sum M_{P_z} = 0$. Similar relations apply with regard to the y and z directions.

The nine components of stress which comprise the *stress tensor* are often described as σ_{ij} where:

$$\sigma_{ij} = \begin{vmatrix} \sigma_{xx} & \sigma_{yx} & \sigma_{zx} \\ \sigma_{xy} & \sigma_{yy} & \sigma_{zy} \\ \sigma_{xz} & \sigma_{yz} & \sigma_{zz} \end{vmatrix}$$

with i and j being iterated over x, y, and z. Since tensor analysis will not be used in this text, the usual tensor notation need not be strictly adhered to. In most instances, normal stresses (tensile or compressive) will be denoted as σ_x, σ_y, σ_z and shear stresses as τ_{xy}, τ_{yz}, τ_{zx}. Note that if $\sum M_p = 0$, $\tau_{xy} = \tau_{yx}$, and so forth, thus the nine components of the stress tensor reduce to six. Conventionally, tensile normal stresses are considered positive while compressive stresses are negative, so all normal stresses in Fig. 1-2 are positive. It

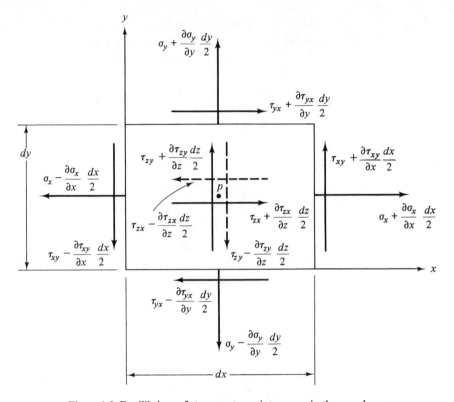

Figure 1-3 Equilibrium of stresses at a point as seen in the *x-y* plane.

is also essential to follow a consistent convention in regard to shear stresses. In this text, positive stresses are *defined* by Fig. 1-2.

If the state of stress varies from point to point this type of change must be accommodated as shown in Fig. 1-3, using the *x-y* plane for reference. Note the following:

1. In moving away from P, any rate of change is assumed to be linear.
2. The sign before any term showing a rate of change does not mean that the change itself is dictated by that sign. For example, the term $\sigma_x + (\partial\sigma_x/\partial x)(dx/2)$ does not imply that the change in σ_x is necessarily increasing as one moves in the positive x direction.
3. Stress components involving σ_z, τ_{yz}, and τ_{xz} are all perpendicular to the *x-y* plane and have no force components parallel to this plane.

Considering a force summation in the x direction, only the terms including σ_x, τ_{yx}, and τ_{zx} are of concern. If the element is in equilibrium, then

$$\sum F_x = 0 = \sum (\text{stress})(\text{area})$$

This produces the following, where the derivation is left as an exercise:

$$\frac{\partial \sigma_x}{\partial x} + \frac{\partial \tau_{yx}}{\partial y} + \frac{\partial \tau_{zx}}{\partial z} = 0 \qquad (1\text{-}1a)^*$$

An identical procedure, using the y and z directions, would produce:

$$\frac{\partial \tau_{xy}}{\partial x} + \frac{\partial \sigma_y}{\partial y} + \frac{\partial \tau_{zy}}{\partial z} = 0 \qquad (1\text{-}1b)$$

$$\frac{\partial \tau_{xz}}{\partial x} + \frac{\partial \tau_{yz}}{\partial y} + \frac{\partial \sigma_z}{\partial z} = 0 \qquad (1\text{-}1c)$$

These are the *equilibrium equations* and although the terms all involve stresses, it should be remembered that they were derived using a force balance.

Three approaches to stress analysis are presented in the rest of this chapter. The first involves an elementary approach to the generalized three-dimensional problem where a known stress state, as related to a particular coordinate system, is transformed to another coordinate system. The second development provides the systematic *transformation* equations which are no more than a series of mathematical expressions that fully encompass the findings of the elementary approach. Finally, because it is such a useful and powerful tool, we devote more attention than usual to the use of Mohr's circle. Although there may appear to be a degree of redundancy in the developments that follow, it is intended that the coverage of stress using different approaches will provide a sound understanding of this important subject.

1-2 A GENERALIZED THREE-DIMENSIONAL ANALYSIS OF A HOMOGENEOUS STRESS STATE

Often, it is important to determine the state of stress on a plane at some angle to those upon which the applied stresses act. Although the state of stress at a point P is not altered, the components of stress acting upon planes other than those which comprise the elemental model will differ. This can be shown most readily by passing an arbitrary plane through the model as shown in Fig. 1-4(a) where the stresses σ_x, τ_{xy}, etc., are known, and the stresses acting upon ABC are to be found in terms of the known stresses.

Considering OP to be normal to ABC, its line of orientation with respect to the x-y-z coordinate system is defined by the three direction cosines shown in Fig. 1-4(b). Let the *total* stress acting on ABC (for the most general analysis) be neither perpendicular nor parallel to this plane; it is indicated as S. The total *force* acting on ABC would be the area of this plane multiplied by

---

*Effects of gravity, acceleration, etc., are of no concern in this book.

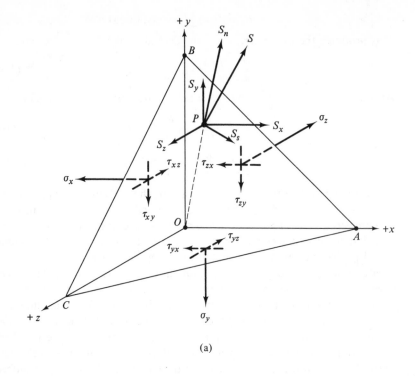

(a)

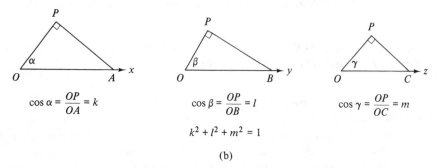

$$\cos \alpha = \frac{OP}{OA} = k \qquad\qquad \cos \beta = \frac{OP}{OB} = l \qquad\qquad \cos \gamma = \frac{OP}{OC} = m$$

$$k^2 + l^2 + m^2 = 1$$

(b)

Figure 1-4 (a) Stresses acting on plane ABC arising from applied stresses shown acting upon three sides of the original stress element; (b) Definition of direction cosines of line OP.

S; this force could then be described by components parallel to the original coordinate system. If each of these force components were then divided by the area of ABC, this would produce stress components S_x, S_y, and S_z as indicated. Because a common *area* is used throughout, we get the following result:

$$S^2 = S_x^2 + S_y^2 + S_z^2 \qquad\qquad (1\text{-}2)$$

If stress components perpendicular and parallel to *ABC* are of greater concern, then using the same concept just discussed, we find:

$$S^2 = S_n^2 + S_s^2 \tag{1-3}$$

where S_n lies along *OP*, is therefore normal to *ABC* and is described by the same direction cosines shown in Fig. 1-4(b). For simplicity, $\cos \alpha = OP/OA = k$, and so forth.

Consider the following symbolization:

$$\text{area of } ABC = N$$
$$\text{area of } OBC = X$$
$$\text{area of } OAC = Y$$
$$\text{area of } OAB = Z$$

If N is considered equal to unity and recalling that the volume of a pyramid is $\frac{1}{3}$(area of the base)(height of the normal to the apex), then

$$(\tfrac{1}{3})X(OA) = (\tfrac{1}{3})N(OP)$$

Rearrangement and substitution leads to $X = k$. It then follows that $Y = l$ and $Z = m$ so the various areas are directly related to the direction cosines defining the line *OP* with respect to the x-y-z system. Recall that $k^2 + l^2 + m^2 = 1$.

Taking a force balance in the x direction,

$$\sum F_x = 0, \qquad S_x(N) = \sigma_x X + \tau_{yx} Y + \tau_{zx} Z$$

or

$$S_x = k\sigma_x + l\tau_{yx} + m\tau_{zx} \tag{1-4}$$

From $\sum F_y$ and $\sum F_z = 0$,

$$S_y = k\tau_{xy} + l\sigma_y + m\tau_{zy}$$
$$S_z = k\tau_{xz} + l\tau_{yz} + m\sigma_z$$

Example 1-1.

An applied stress state is described by:

$$\sigma_{ij} = \begin{vmatrix} 20 & 3 & 8 \\ 3 & 15 & 5 \\ 8 & 5 & 10 \end{vmatrix}$$

Note that all stresses are indicated as *positive* according to the convention described by Fig. 1-2. With reference to Fig. 1-4, a plane such as *ABC* has a normal whose direction angles are given by $\alpha = 45°$, $\beta = 50°$ and $\gamma = 72.8°$ (i.e., $k = 0.707$, $l = 0.643$, and $m = 0.296$).

Determine the magnitudes of the total state of stress, *S*, and its normal component, S_n, acting on *ABC*.

Solution.

Using Eq. (1-4):

$$S_x = 20(0.707) + 3(0.643) + 8(0.296) = 18.44$$
$$S_y = 3(0.707) + 15(0.643) + 5(0.296) = 13.25$$
$$S_z = 8(0.707) + 5(0.643) + 10(0.296) = 11.83$$

From Eq. (1-2):

$$S^2 = S_x^2 + S_y^2 + S_z^2 = 655.55, \qquad S = 25.6$$
$$S_n = kS_x + lS_y + mS_z = 25.05 \qquad \text{[This can be checked with Eq. (1-5).]}$$

Equations (1-4) give the stresses S_x, S_y, and S_z as functions of the direction cosines and the known stresses, σ_x, etc.

If greater interest involves the components S_n and S_s, vector analysis gives $S_n = kS_x + lS_y + mS_z$, and by substituting Eq. (1-4) the result is

$$S_n = k^2\sigma_x + l^2\sigma_y + m^2\sigma_z + 2(kl\tau_{yx} + lm\tau_{yz} + mk\tau_{zx}) \qquad (1\text{-}5)$$

Thus S_n is determined from the direction cosines and the known stresses. Using Eqs. (1-2) and (1-3) gives

$$S_s^2 = (S_x^2 + S_y^2 + S_z^2) - S_n^2 \qquad (1\text{-}6)$$

If Eqs. (1-4) and (1-5) were substituted into (1-6), the result would give S_s in terms of the known stresses and direction cosines. The reader can conduct this as an exercise if desired; the algebraic manipulation is lengthy.

The above developments portray the state of stress on an arbitrary plane with reference to the original coordinate system and are intended to indicate the three-dimensional nature of stress as a general consequence.

As a further point, suppose the plane ABC was oriented such that S was perpendicular to it. In essence, no shear component acts on ABC and, of course, the direction cosines defining the line from the origin, O, that is now normal to ABC would differ in magnitude from those in the previous analysis. Symbolically, however, k, l, and m can still be used. With this in mind,

$$S_x = kS, \qquad S_y = lS, \qquad \text{and} \qquad S_z = mS$$

These three relationships, if substituted into Eq. (1-4), produce the following:

$$k(\sigma_x - S) + l(\tau_{yx}) + m(\tau_{zx}) = 0$$
$$k(\tau_{xy}) + l(\sigma_y - S) + m(\tau_{zy}) = 0 \qquad (1\text{-}7)$$
$$k(\tau_{xz}) + l(\tau_{yz}) + m(\sigma_z - S) = 0$$

These three homogeneous equations give real roots other than zero only if the determinant is zero. Setting the determinant to zero and expanding gives a cubic equation whose three roots are the three *principal stresses* (i.e., the stresses on planes of zero shear stress). Denoting the stress S then as σ_p gives:

$$\sigma_p^3 - (\sigma_x + \sigma_y + \sigma_z)\sigma_p^2 - (\tau_{xy}^2 + \tau_{yz}^2 + \tau_{zx}^2 - \sigma_x\sigma_y - \sigma_y\sigma_z$$

$$- \sigma_z\sigma_x)\sigma_p - (\sigma_x\sigma_y\sigma_z + 2\tau_{xy}\tau_{yz}\tau_{zx} - \sigma_x\tau_{yz}^2 - \sigma_y\tau_{zx}^2 - \sigma_z\tau_{xy}^2) = 0 \quad (1\text{-}8)$$

Regardless of the coordinate system chosen, these three roots or principal stresses are constant for a given state of stress. In essence, the coefficients in parentheses must not *vary* with a change in coordinate system. They are, therefore, referred to as the invariants of the stress tensor, and are denoted as follows:

INVARIANTS

$$\text{First} \equiv I_1 = (\sigma_x + \sigma_y + \sigma_z)$$
$$\text{Second} \equiv I_2 = (\tau_{xy}^2 + \tau_{yz}^2 + \tau_{zx}^2 - \sigma_x\sigma_y - \sigma_y\sigma_z - \sigma_z\sigma_x) \quad (1\text{-}9)$$
$$\text{Third} \equiv I_3 = (\sigma_x\sigma_y\sigma_z + 2\tau_{xy}\tau_{yz}\tau_{zx} - \sigma_x\tau_{yz}^2 - \sigma_y\tau_{zx}^2 - \sigma_z\tau_{xy}^2)$$

Example 1-2.

A given stress state is expressed as

$$\sigma_{ij} = \begin{vmatrix} 4 & 2 & 3 \\ 2 & 6 & 1 \\ 3 & 1 & 5 \end{vmatrix}$$

The units of each stress are in MPa and all stresses are denoted as positive as defined in Fig. 1-2. Find the magnitudes of the principal stresses and the direction cosines defining the line of action of the largest principal stress with respect to the original x-y-z coordinate system.

Solution.

Equation (1-8) can be written as:

$$\sigma_p^3 - I_1\sigma_p^2 - I_2\sigma_p - I_3 = 0$$

where the invariants are defined by Eq. (1-9). Note that care must be exercised with respect to signs since different texts may use modified forms of the invariants.

Here,

$$I_1 = 4 + 6 + 5 = +15$$
$$I_2 = 4 + 1 + 9 - 24 - 30 - 20 = -60$$
$$I_3 = 120 + 12 - 4 - 54 - 20 = +54$$

so,

$$\sigma_p^3 - 15\sigma_p^2 + 60\sigma_p - 54 = 0$$

Although mathematical techniques for solving cubic equations exist (and certainly computer solutions are available), many hand calculators can be programmed to produce such a solution. As a last resort, sensible trial and error can always be used. The three roots of the above equation are:

$$\sigma_1 = +9, \quad \sigma_2 = +4.73, \quad \text{and} \quad \sigma_3 = +1.27$$

To obtain the direction cosines of the largest principal stress, the set in Eq. (1-7) is first used. Introducing the pertinent stress magnitudes, we get:

$$k(4 - 9) + l(2) + m(3) = 0$$
$$k(2) + l(6 - 9) + m(1) = 0$$
$$k(3) + l(1) + m(5 - 9) = 0$$

or

$$-5k + 2l + 3m = 0$$
$$2k - 3l + 1m = 0$$
$$3k + 1l - 4m = 0$$

Using any two of the above three equations plus the identity that $k^2 + l^2 + m^2 = 1$, the values for k, l, and m are then found. For this *particular* example, $k = l = m$ so that the line of action of the principal stress σ_1, whose magnitude is 9 MPa would be equally inclined with respect to the original x-y-z coordinate system at an angle of 54.74 degrees. This comes from:

$$k^2 + l^2 + m^2 = 1 = 3k^2 = 3l^2 = 3m^2$$

so,

$$k = \left(\frac{1}{3}\right)^{1/2}$$

and $\cos \alpha = k$, so $\alpha = 54.74$ degrees $= \beta = \gamma$

Example 1-3.

Consider the applied state of stress and values of the three principal stresses found in Example 1-2. Determine the invariants using the principal stresses and compare them with the magnitudes found using the applied stresses.

Solution.

The invariants determined in Example 1-2 were:

$$I_1 = +15, \qquad I_2 = -60, \qquad I_3 = +54$$

If the principal stresses are used, then by definition, no shear stresses act on such planes, so the invariants become:

$$I_1 = \sigma_1 + \sigma_2 + \sigma_3 = 9 + 4.73 + 1.27 = +15$$
$$I_2 = -(\sigma_1\sigma_2 + \sigma_2\sigma_3 + \sigma_3\sigma_1) = -(42.57 + 6.00 + 11.43) = -60$$
$$I_3 = (\sigma_1\sigma_2\sigma_3) = +54$$

which are identical to the earlier values as they must be.

The convention used in this book for defining positive and negative stresses was mentioned in connection with Fig. 1-2 and the need for such a convention can be now explained in a more obvious manner. Regarding Eq. (1-8), the *magnitude* of the three invariants (and, therefore, the magnitudes of the three principal stresses) will certainly depend upon the algebraic values of the individual stresses; this is shown in Example 1-2. Now it should be obvious that if two stress states are identical except for the sign of one normal stress, say σ_x, the magnitude of each invariant will differ for these two situations since σ_x enters as a term to the first power. Thus a value of $+10$ compared to -10 for σ_x will lead to different values of the three roots or principal stresses.

This result may not be as obvious where shear stresses are considered. Suppose that two stress states are identical except that one shear stress, say τ_{xy}, is $+5$ in one case and -5 in the second, again using the convention described by Fig. 1-2. Here, only the magnitude of the third invariant will differ for these two situations since it is the only quantity that involves τ_{xy} to the first power (terms involving τ_{xy}^2 will be unaffected by the *sign* of τ_{xy}). In effect, the sign of shear stresses must be handled properly since the magnitudes of the principal stresses would again be different for these two situations. Although any convention for the designation of positive and negative stresses is arbitrary, once it is defined it must be followed with consistency.

1-3 STRESS TRANSFORMATIONS

To provide a systematic approach to the transformation of stress from one coordinate system to another, a generalized set of equations can be developed. The reader is undoubtedly familiar with analogous transformations for vector quantities such as force. Consider the situation shown in Fig. 1-5(a) where

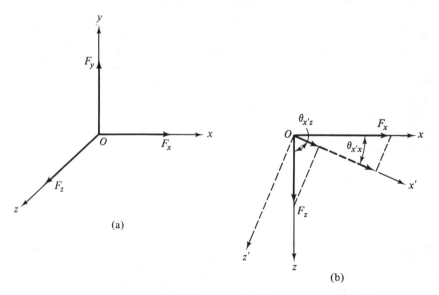

Figure 1-5 (a) Forces related to Cartesian coordinates; (b) Transformation of forces from an x-y-z system of coordinates to an x'-y-z' system.

forces F_x, F_y, and F_z act along the x, y, z reference axes. Now with respect to a new set of reference axes, we consider the force components that result. For simplicity, assume y' is the same as y such that the new x' and z' axes are in the same plane as x and z. The view normal to this plane (i.e., along the y axis) is shown in Fig. 1-5(b). The force component $F_{x'}$ is composed of the projections of F_x and F_z on the x' axis, thus,

$$F_{x'} = F_x \cos \theta_{x'x} + F_z \cos \theta_{x'z}$$

or

$$F_{x'} = kF_x + mF_z$$

in the earlier notation of direction cosines ($k = \cos \theta_{x'x}$ etc.). It should be obvious that if the y' axis were not parallel to the y axis, the force F_y would also contribute to $F_{x'}$ with the result:

$$F_{x'} = kF_x + lF_y + mF_z \tag{1-10}$$

Similar relationships could be developed for $F_{y'}$ and $F_{z'}$ using the *proper set* of direction cosines for each transformation. In the shorthand symbolization of tensor notation the type of transformation just discussed is expressed as:

$$F_i = \sum l_{ij}F_j, \qquad j = x, y, z, \qquad i = x', y', z', \qquad \text{and} \qquad l_{ij} = \cos \theta_{ij}$$

Note that vector quantities such as forces involve single sets of direction cosines for transformation of axes so that only one symbol (l_{ij}) appears under the summation sign above.

For the transformation of tensor quantities such as stress, first consider Fig. 1-6(a) where the uniaxial tensile stress, σ_{zz}, is imposed and transformation from the x, y, z coordinate system to the x, y', z' axes is desired. (Note x and x' are coincident for simplicity.) Looking down the x axis, Fig. 1-6(b) shows the result. The *force* in the z direction is $F_z = \sigma_{zz}A_z$, A_z being the area normal to z. Thus, $F_{z'} = F_z \cos \theta_{z'z}$ is the component of F_z acting along the z' axis. The area $A_{z'}$ which is normal to z' is:

$$A_{z'} = \frac{A_z}{\cos \theta_{z'z}}$$

Therefore,

$$\sigma_{z'z'} = \frac{F_{z'}}{A_{z'}} = \frac{F_z \cos \theta_{z'z}}{\dfrac{A_z}{\cos \theta_{z'z}}}$$

or

$$\sigma_{z'z'} = \sigma_{zz} \cos^2 \theta_{z'z}$$

For the fully generalized case, this type of transformation is expressed as

$$\sigma_{ij} = \sum l_{im}l_{jn}\sigma_{mn}$$

with m, n iterated over x, y, z and i, j iterated over x', y', z'. Thus, the *complete* expression for $\sigma_{x'x'}$ becomes:

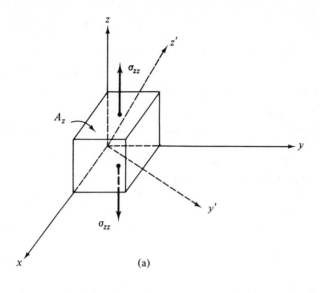

(a)

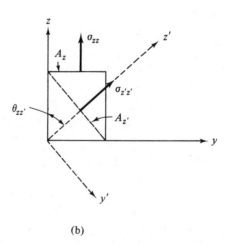

(b)

Figure 1-6 (a) Uniaxial stress defined by an x-y-z system; (b) transformation of stress σ_{zz} to an x-y'-z' system.

$$\sigma_{x'x'} = l_{x'x}l_{x'x}\sigma_{xx} + l_{x'y}l_{x'y}\sigma_{yy} + l_{x'z}l_{x'z}\sigma_{zz}$$
$$+ l_{x'x}l_{x'y}\sigma_{xy} + l_{x'y}l_{x'z}\sigma_{yz} + l_{x'z}l_{x'x}\sigma_{zx}$$
$$+ l_{x'y}l_{x'x}\sigma_{yx} + l_{x'z}l_{x'y}\sigma_{zy} + l_{x'x}l_{x'z}\sigma_{xz} \qquad (1\text{-}11)$$

By appropriate interchange of subscripts, equivalent expressions could be developed for $\sigma_{y'y'}$, $\sigma_{z'z'}$, $\sigma_{x'y'}$, etc.* From previous observations regarding

*See, for example, N. H. Polakowski and E. J. Ripling, *Strength and Structure of Engineering Materials* (Englewood Cliffs, N.J.: Prentice-Hall, Inc., 1966) pp. 32–5.

complementary shear ($\sigma_{ij} = \sigma_{ji}$ or $\tau_{ij} = \tau_{ji}$) and using the simplified notation for normal stresses ($\sigma_{ii} = \sigma_i$), Eq. (1-11) could then be shortened to:

$$\sigma_{x'} = l^2_{x'x}\sigma_x + l^2_{x'y}\sigma_y + l^2_{x'z}\sigma_z + 2(l_{x'x}l_{x'y}\tau_{xy} + l_{x'x}l_{x'z}\tau_{xz} + l_{x'y}l_{x'z}\tau_{yz}) \quad (1\text{-}12)$$

Note the equivalence of Eq. (1-12) to Eq. (1-5).

From the above developments it can be seen that stress transformations require the general use of two sets of direction cosines whereas vector transformations require only one set. It should be apparent that in situations involving lengthy mathematical developments, the shorthand type of tensor notation is extremely efficient and compact, but for most problems that will be presented in this text, the type of transformations encountered will be simplifications of the full expressions since some of the stress components will be zero.

Example 1-4.

Using the data given in Example 1-1, find the magnitude of the stress acting normal to ABC using Eq. (1-12).

Solution.

Here, $\sigma_{x'}$ is the normal stress acting in the x' direction which is defined by the angles $\alpha = 45°$, $\beta = 50°$ and $\gamma = 72.8°$. Thus,

$$l_{x'x} = \cos 45° = 0.707 \quad \text{(this is } k\text{)}$$
$$l_{x'y} = \cos 50° = 0.643 \quad \text{(this is } l\text{)}$$
$$l_{x'z} = \cos 72.8° = 0.296 \quad \text{(this is } m\text{)}$$
$$\sigma_{x'} = 20(0.707)^2 + 15(0.643)^2 + 10(0.296)^2 + 2[(0.707)(0.643)(3) + (0.707)(0.296)(8) + (0.643)(0.296)(5)]$$
$$\sigma_{x'} = 25.05$$

which is S_n in Example 1-1.

Note: Students are often confused about the handling of the "signs" of stress components; the use of a consistent convention is designed to avoid such concerns. The convention used in most texts is followed here and is described by Fig. 1-2, where all 9 components of the stress tensor are defined as being positive.

Consider σ_{xx} first. The first subscript indicates the plane whose normal acts in the positive x direction; thus, this may be viewed as a "positive" plane. The second subscript indicates the direction in which the stress acts; here this is in the positive x direction. Two such "positive" components in combination therefore define a positive stress and such normal stresses are tensile by definition.

Now consider σ_{xy}. Again this acts on a positive x plane (i.e., its normal is in the positive x direction) and the stress itself acts in the positive y direction. Thus this shear stress, σ_{xy} is defined as positive. Similar arguments show that all other components in Fig. 1-2 are positive according to this convention.

If the direction of either σ_{xx} or σ_{xy} were reversed on Fig. 1-2, they would still act on a positive x plane but in negative directions. The combination of positive and negative components therefore defines "negative" stresses. Thus the negative normal stress, σ_{xx}, would be compressive whereas σ_{xy} would be a negative shear stress and both should be preceded by negative signs when written in the tensor form given in Example 1-1. It is essential that such negative signs be *included* in the type of calculations performed in all previous examples!

Finally, if reference is made to Fig. 1-4, the stress shown as σ_x (really σ_{xx}) acts on a plane whose normal is in the negative x direction; the direction of the stress is also shown acting in the negative direction. When a combination of two negative components occurs, the stress is again considered positive. Since a tensile stress is shown, this provides consistency with such a convention. Note that *all* other stresses in Fig. 1-4 have negative–negative combinations so all such stresses are defined as being positive.

1-4 MOHR'S CIRCLE—ITS USE AND LIMITATION

It is important at the outset to realize that *every* state of stress can be considered as three-dimensional. In the case of uniaxial tension, two of the principal stresses happen to be zero while with *plane stress* (biaxial state of stress), one of the principal stresses is zero. Each of these is a specialized case of the more general three-dimensional situation. Figure 1-7 describes these two cases.

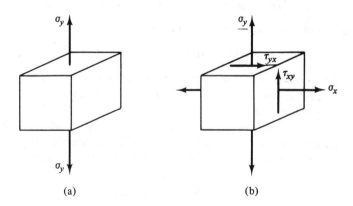

(a)   (b)

Figure 1-7 Stress elements for (a) uniaxial and (b) biaxial states.

In Fig. 1-7, σ_x and σ_z are principal stress in the uniaxial situation while σ_z is a principal stress in the biaxial case. This is immediately obvious because no shear stresses act on those planes, and, in essence, one or two *roots* of Eq. (1-8) are known immediately. For any stress state that is equivalent to a biaxial or uniaxial condition, Mohr's circle can be applied directly. The *prime requirement* is that *shear stresses* must be absent on at least one plane, thus, that plane is a principal plane and the normal stress (whether tensile, compressive, or zero) is a principal stress. Then, the other two principal stresses (or unknown roots of the cubic equation) must lie in the plane perpendicular to the known principal direction.

Certainly, the substitution of the known root into Eq. (1-8) would reduce the cubic to a quadratic which could be easily solved. Mohr's* circle is a graphical plot of that reduced equation.

Consider the biaxial stress state shown in Fig. 1-8(a), where for simplicity $\sigma_z = \tau_{xz} = \tau_{yz} = 0$. It was shown earlier that as the plane of interest was rotated, the state of stress changed. Here, θ is used to denote some arbitrary angle with respect to an initially arbitrary x-y-z coordinate system. Note that with Eq. (1-11) the stress state on this "θ plane" could be found directly. For cases of the type being considered, the use of Mohr's circle does the same thing; in fact, Mohr's circle is a *graphical tensor transformation*! Before discussing this further, consider the analysis that follows.

Regarding Fig. 1-8(b), equilibrium of forces along the x' and y' axes must be zero.

Considering $\sum F_{x'} = 0$ we get the result:

$$\sigma_{x'} = \sigma_x \cos^2 \theta + \sigma_y \sin^2 \theta + 2\tau_{xy} \sin \theta \cos \theta \qquad (1\text{-}13)$$

With the aid of trigonometric identities,† Eq. (1-13) may be converted to:

$$\sigma_{x'} = \tfrac{1}{2}(\sigma_x + \sigma_y) + \tfrac{1}{2}(\sigma_x - \sigma_y) \cos 2\theta + \tau_{xy} \sin 2\theta \qquad (1\text{-}14)$$

The force balance in the y' direction leads to:

$$\tau_{x'y'} = \tfrac{1}{2}(\sigma_x - \sigma_y) \sin 2\theta - \tau_{xy} \cos 2\theta \qquad (1\text{-}15)‡$$

Interest often centers on those planes where $\tau_{x'y'}$ is zero (i.e., the principal planes) or where $\tau_{x'y'}$ is a maximum. The latter is of great interest when the onset of plastic flow is of concern.

In order to find the two principal stresses in the x-y plane, $\tau_{x'y'}$ in Eq. (1-15) is set equal to zero; the unique value of θ that defines the principal directions with respect to the original coordinate system is then given by:

$$\tan 2\theta = \frac{2\tau_{xy}}{\sigma_x - \sigma_y} \qquad (1\text{-}16)$$

Recall that since this is a biaxial situation, the third principal direction is perpendicular to the x-y plane. From Eq. (1-16),

---

*O. Mohr, *Zivilingeneur*, 1882, p. 113.

†$\sin 2\theta = 2 \sin \theta \cos \theta$ and $\cos 2\theta = 2 \cos^2 \theta - 1 = 1 - 2 \sin^2 \theta$.

‡If the direction of $\tau_{x'y'}$ is taken opposite to that shown on Fig. 1-8, then Eq. (1-15) becomes $\tau_{x'y'} = -(\sigma_x - \sigma_y) \sin 2\theta/2 + \tau_{xy} \cos 2\theta$. This would provide direct consistency with Eq. (1-11) yet most sources give Eq. (1-15) as shown. Example 1-5 and the accompanying explanation is intended to clarify this point.

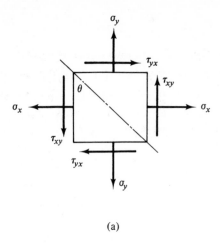

(a)

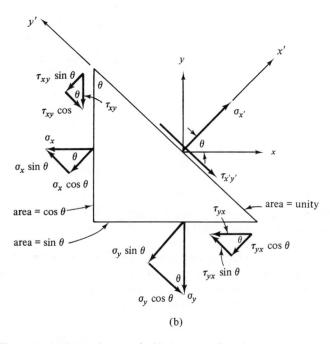

(b)

Figure 1-8 (a) Stress element of a biaxial state of stress; (b) basis for force equilibrium analysis.

$$\sin 2\theta = \frac{2\tau_{xy}}{[(\sigma_x - \sigma_y)^2 + 4\tau_{xy}^2]^{1/2}}$$

and

$$\cos 2\theta = \frac{(\sigma_x - \sigma_y)}{[(\sigma_x - \sigma_y)^2 + 4\tau_{xy}^2]^{1/2}} \tag{1-17}$$

17

Substitution of these findings into Eq. (1-14) gives:

$$(\sigma_{x'})_p = \tfrac{1}{2}(\sigma_x + \sigma_y) \pm \tfrac{1}{2}[(\sigma_x - \sigma_y)^2 + 4\tau_{xy}^2]^{1/2} = A \pm B \qquad (1\text{-}18)$$

These two expressions give the principal stresses in the x-y plane and from this point on such stresses will be denoted as σ_1, σ_2, and σ_3. From Eq. (1-18), $\sigma_1 = A + B$ and $\sigma_2 = A - B$ while σ_3 is known to act in the third (z) direction.

With regard to the planes where $\tau_{x'y'}$ is maximum (i.e., τ_{max}), set $(d\tau_{x'y'})/d\theta = 0$ in Eq. (1-15). This leads to:

$$\tan 2\theta = \frac{-(\sigma_x - \sigma_y)}{2\tau_{xy}} \qquad (1\text{-}19)$$

Using expressions similar to Eq. (1-17) and substituting into Eq. (1-15) gives:

$$\tau_{max} = \tfrac{1}{2}[(\sigma_x - \sigma_y)^2 + 4\tau_{xy}^2]^{1/2} \qquad (1\text{-}20)$$

Two points are worth noting here. First, Eq. (1-20) defines the largest shear stress in the x-y plane. An even *larger* shear stress might act in one of the other two planes; appropriate examples will make this evident. It is also noted that comparisons of Eqs. (1-16) and (1-19) show that τ_{max} acts on planes oriented 45° from those exposed to the principal stresses.

Example 1-5.

A stress state is described by:

$$\sigma_{ij} = \begin{vmatrix} 20 & -4 & 0 \\ -4 & -15 & 0 \\ 0 & 0 & 10 \end{vmatrix}$$

Since the stress given as $\sigma_z = 10$ is a principal stress, Eqs. (1-14), (1-15), (1-18), and (1-20) are applicable for this problem and reference to Fig. 1-8 is useful.

(a) Consider a plane located at an angle of $\theta = 30°$ as indicated in Fig. 1-8. Determine the normal and shear stresses acting on that plane.

(b) Determine the magnitudes of the three principal stresses and the largest shear stress acting in the x-y plane.

Solution.

(a) From Eqs. (1-14) and (1-15):

$$\sigma_{x'} = \tfrac{1}{2}(20 - 15) + \tfrac{1}{2}[20 - (-15)] \cos 60° + (-4) \sin 60°$$
$$= 2.5 + \tfrac{1}{2}(35)(0.5) - 4(0.866)$$
$$= 7.79 \quad (+, \text{ therefore, tensile})$$
$$\tau_{x'y'} = \tfrac{1}{2}[20 - (-15)] \sin 60 - (-4) \cos 60$$
$$= \tfrac{1}{2}(35)(0.866) + 4(0.5)$$
$$= 17.16$$

A word of caution is appropriate here. Both $\sigma_{x'}$ and $\tau_{x'y'}$ are indicated as positive stresses in the solution above. Since Eqs. (1-14) and (1-15) were derived from a force balance using Fig. 1-8, any stress state that produces positive values for $\sigma_{x'}$ and $\tau_{x'y'}$ using Eqs. (1-14) and (1-15) implies that these calculated stresses act in the direction shown on Fig. 1-8. In such instances, $\tau_{x'y'}$ is considered *negative* in regard to the tensor convention discussed after Example 1-4 and shown in Fig. 1-2. Using Eq. (1-11) will also give a negative value for $\tau_{x'y'}$ in this example!

(b) Using Eq. (1-18):

$$\sigma_p = \tfrac{1}{2}(20 - 15) \pm \tfrac{1}{2}\{[20 - (-15)]^2 + 4(-4)^2\}^{1/2}$$
$$= 2.5 \pm \tfrac{1}{2}(35.90) = 2.5 \pm 17.95$$

so

$$\sigma_1 = 20.45, \qquad \sigma_2 = 10 \quad (\text{i.e., } \sigma_z), \qquad \sigma_3 = -15.45$$

With Eq. (1-20), $\tau_{\max} = 17.95$

The previous developments have made no mention of a plot of Mohr's circle but the following will indicate the basis behind this concept.

Rearrange Eqs. (1-14) and (1-15) to give:

$$[\sigma_{x'} - \tfrac{1}{2}(\sigma_x + \sigma_y)] = \tfrac{1}{2}(\sigma_x - \sigma_y)\cos 2\theta + \tau_{xy}\sin 2\theta$$
$$\tau_{x'y'} = \tfrac{1}{2}(\sigma_x - \sigma_y)\sin 2\theta - \tau_{xy}\cos 2\theta$$

Squaring the above equations, then adding, gives:

$$\left[\sigma_{x'} - \frac{1}{2}(\sigma_x + \sigma_y)\right]^2 + \tau_{x'y'}^2 = \left(\frac{\sigma_x - \sigma_y}{2}\right)^2 + \tau_{xy}^2 \qquad (1\text{-}21)$$

This is the equation of a circle whose center is at $[\tfrac{1}{2}(\sigma_x + \sigma_y), 0]$ and whose radius is $[(\sigma_x - \sigma_y)^2/4 + \tau_{xy}^2]^{1/2}$. The concept behind Mohr's circle should now be evident. Every point *on the circle* defines the stress state acting on planes at any angle θ from the original x or y axis.

For the correct construction of a Mohr's circle, certain rules are followed and a consistent handling of positive and negative stresses is essential. It should be pointed out that such care is required only if the proper *orientation* of planes is desired. No such concern is needed if only the magnitudes of the principal stresses are sought. This will be shown shortly. Although various conventions are in use, the following is recommended:

1. Normal stresses are plotted to scale along the abscissa with tensile stresses considered positive and compressive negative.

2. Shear stresses are plotted along the ordinate to the same scale as used for normal stresses. A shear stress that would tend to cause *clockwise* rotation of the stress element in the physical plane is considered positive while negative shear stresses tend to cause counterclockwise rotation.*

---

*For the construction of Mohr's circle and *only for this construction*, the sign convention for shear stresses related to Fig. 1-2 is abandoned.

3. Angles between lines of direction on the Mohr plot are twice the indicated angles on the physical plane. Recall that the angle θ on Fig. 1-8 appears as 2θ in all of the equations.

To illustrate the above comments, Fig. 1-9(a) shows the physical plane and *b* the corresponding Mohr's circle. The point having coordinate values of σ_x and $-\tau_{xy}$ (counterclockwise) is shown as *H*, while *K* has coordinates of

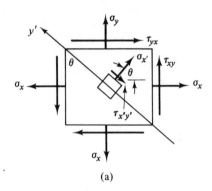

(a)

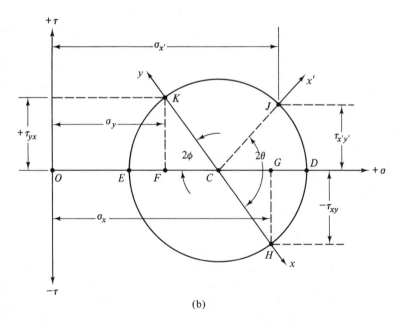

(b)

Figure 1-9 Stress element (a) and Mohr's circle (b) for analyzing stresses in the *x-y* plane.

σ_y and $+\tau_{yx}$. (Again, recall that $\tau_{xy} = \tau_{yx}$.) The line yx drawn through H and K cuts the abscissa at C which is the *center* of the circle whose radius is CK or CH. Since θ is counterclockwise with respect to the x axis in Fig. 1-9(a) a value of 2θ is produced in a counterclockwise direction from the x axis on the circle plot (i.e., from CH) as shown. The intersection at point J then gives the magnitudes of $\sigma_{x'}$ and $\tau_{x'y'}$ since they are now scaled from the plot. The values of OD and OE provide the magnitudes of the principal stresses σ_1 and σ_2 (τ is zero at these points) while 2ϕ is angle between the x direction and principal direction 1. Note that σ_1 is acting counterclockwise from σ_x and reference to Fig. 1-9(a) will show that this makes physical sense. To ensure clarity on this point, several illustrations are presented.

Suppose both shear stresses in Fig. 1-9(a) were considered positive as shown by points K and H' in Fig. 1-10. The perpendicular bisector of KH'

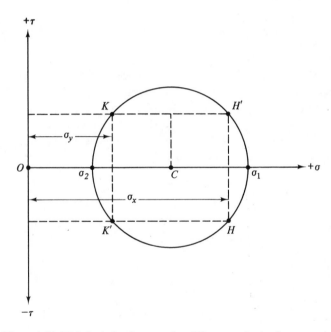

Figure 1-10 Mohr's circle of stress using different methods of construction.

leads to the location of C as center and with CK as radius, one obtains the same circle as in Fig. 1-9(b), thereby producing the same *magnitudes* of σ_1 and σ_2. Alternatively, if both shear stresses were assumed negative, K' and H would be located as shown; again, the magnitudes of σ_1 and σ_2 would be defined correctly.

Since, however, the correct *orientation* of σ_1 and σ_2 with respect to the original coordinate system is often of importance, a systematic convention that assigns positive and negative signs to the shear stresses will lead to proper orientation. With the approach that lead to Fig. 1-9, the stress element involving principal stresses is shown in Fig. 1-11. This indicates that σ_1 acts along a line that is counterclockwise by ϕ degrees from the original x axis.

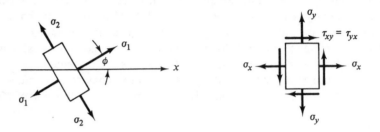

Figure 1-11 Principal stress element of correct orientation with respect to an element showing applied stresses.

Suppose that the opposite shear sign notation had been used (i.e., clockwise is negative). Then the connection between H' and K' (Fig. 1-10) would produce C and the same circle. However, the x axis would now be related to H' and the angle, 2ϕ, between CH' and the abscissa would indicate that σ_1 acts along a line that is *clockwise* from the x axis as shown below in Fig. 1-12. Now physically, due to the shear directions given in Fig. 1-9(a) it would

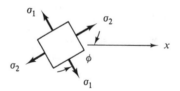

Figure 1-12 Principal stress element of incorrect orientation as related to applied stresses in Fig. 1-11.

be expected that the stress element would tend to distort into a diamond shape pointing in the direction of the shears as indicated in Fig. 1-13.

It can be rationalized that the same distortion could be produced by principal stresses shown acting in Fig. 1-13; thus the largest principal stress must logically act in the direction shown. Obviously such physical consistency does not follow if σ_1 acts as shown in Fig. 1-12 so the need for a consistent convention when plotting a Mohr's circle is evident. Note that the convention

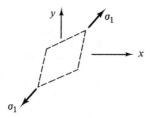

Figure 1-13 Distortion of stress element due to principal stresses.

suggested in this discussion is not the only one employed. It is merely a matter of personal preference and appears to be most widely used.

Returning to Fig. 1-9, the following useful relationships may be noted:

1. The center of the circle is located at point C and $OC = OE + EC$ which equals $[\frac{1}{2}(\sigma_1 + \sigma_2)]$ or $[\frac{1}{2}(\sigma_x + \sigma_y)]$.
2. The radius of the circle $(R) = CD = [\frac{1}{2}(\sigma_1 - \sigma_2)]$ or $R = (KF^2 + FC^2)^{1/2}$ so $R = [(\sigma_x - \sigma_y)^2/4 + \tau_{xy}^2]^{1/2} = \tau_{xy}/\sin 2\phi$ with $\tan 2\phi = [2\tau_{xy}/(\sigma_x - \sigma_y)]$ (see Eq. (1-20)).
3. $(\sigma_1 - \sigma_2) = 2\tau_{max} = [(\sigma_x - \sigma_y)^2 + 4\tau_{xy}^2]^{1/2} = 2R$.
4. $\sigma_1 = OD = OF + FC + CD = [\frac{1}{2}(\sigma_x + \sigma_y)] + \tau_{max}$ (see Eq. (1-18)).
5. $\sigma_2 = OE = OF + FC - EC = [\frac{1}{2}(\sigma_x + \sigma_y)] - \tau_{max}$ (see Eq. (1-18)).
6. $\tau_{xy} = R \sin 2\phi = [\frac{1}{2}(\sigma_1 - \sigma_2)] \sin 2\phi$.

Example 1-6.

Repeat Example 1-5 using a Mohr's circle construction.

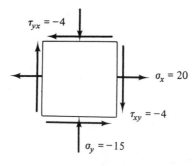

Fig. E1-6

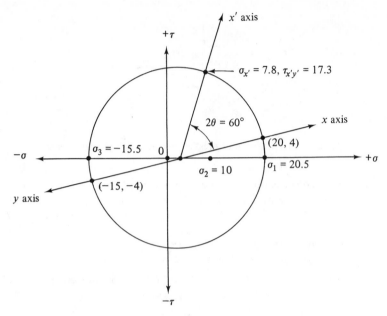

Fig. E1-6 (cont.)

Solution.

From the stress tensor notation, τ_{xy} and τ_{yx} are negative so they act as shown in the sketch. For the construction by Mohr's circle, τ_{xy} acts *clockwise* and is defined as positive whereas τ_{yx} is negative (See Fig. E1-6).

Now suppose the stress element is subjected to principal stresses as shown in Fig. 1-14(a) and leads to the circle plot in 1-14(b). Regarding the stresses on an arbitrary plane oriented θ degrees counterclockwise from the "1" direction, note the following:

1. $\sigma_\theta = OB = OA + AB = [\tfrac{1}{2}(\sigma_1 + \sigma_2)] + [\tfrac{1}{2}(\sigma_1 - \sigma_2)] \cos 2\theta$
    which agrees with Eq. (1-14) since τ_{xy} is zero here.
2. $\tau_\theta = CB = [\tfrac{1}{2}(\sigma_1 - \sigma_2)] \sin 2\theta$ which agrees with Eq. (1-15).

A few common stress states are shown below:

(a) Uniaxial tension (prior to necking) where $\sigma_x \neq 0$ but all other applied stresses are zero, thus $\sigma_x = \sigma_1$ (Fig. 1-15).
(b) Uniform direct compression would give a mirror image of Fig. 1-15 where the compressive stress σ_x has the smallest *algebraic* value; thus $\sigma_x = \sigma_3$, $\sigma_1 = \sigma_2 = 0$. This introduces a convention

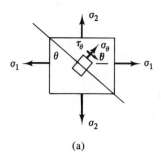

(a)

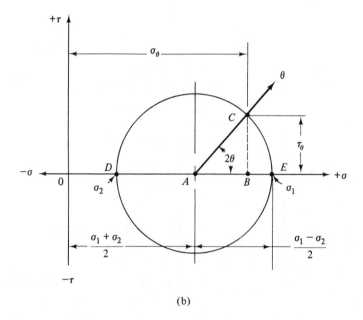

(b)

Figure 1-14 Principal stress element (a) and corresponding Mohr's circle (b).

that $\sigma_1 > \sigma_2 > \sigma_3$ which will be used *wherever* feasible and implies that a stress of zero is greater than a compressive stress.

(c) A thin sheet subjected to tensile loads parallel to the plane of the sheet such that $\sigma_1 > \sigma_2, \sigma_3 = 0$. This loading is referred to as unbalanced biaxial tension and is shown in Fig. 1-16. Note that the single largest shear stress is described by the radius of the circle bounded by σ_1 and σ_3. This finding will *always* result when the principal stresses determined by the circle plot describing the

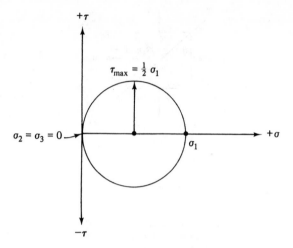

Figure 1-15 Mohr's circle for uniaxial tension

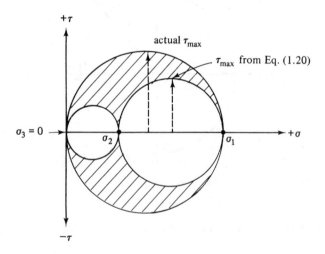

Figure 1-16 Mohr's circle for unbalanced biaxial tension.

known stress state are both *larger* than the third principal stress which must be known at the outset.

In regard to Fig. 1-16 a stress state for any arbitrary orientation of axes must lie *on* one of the three circles or *within* the shaded area.

(d) Now consider the stress state as shown in Fig. 1-17. Here, one principal stress, σ_3, is known to be compressive. Looking along

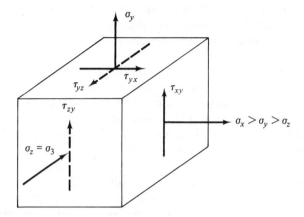

Figure 1-17 Generalized stress element where one principal stress is known.

the z axis means the other two principal stresses must lie in the x-y plane and a typical general plot is shown in Fig. 1-18.

However, if a τ_{zy} were present (shown dotted), σ_z is not a principal stress and it is not possible to determine the principal stresses from a plot of Mohr's circle. Instead the cubic equation

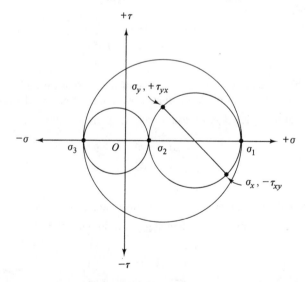

Figure 1-18 Mohr's circle for the three-dimensional stress state shown in Fig. 1-17.

must be solved for the three roots. In most instances when a circle can be correctly plotted, its major use is to determine the two unknown principal stresses.

One last point is of interest here. As discussed in connection with Fig. 1-10, the sign of the shear stresses must be handled properly in a plot of Mohr's circle if stress orientations are of concern; no such concern is necessary in regard to the *magnitudes* of the two principal stresses found from the plot. Such was not the case in the discussion following Example 1-3 where a three-dimensional stress state existed; that is, it was essential to handle the signs of shear stresses correctly since the *magnitudes* of the principal stresses were dependent upon such procedures. This apparent anomaly can best be explained by considering Eq. (1-8). For cases where Mohr's circle can be applied, the term involving the *product* of the shear stresses goes to zero, in fact, only one shear stress can be nonzero. Since such a stress enters as a squared term in the second and third invariants, its *sign* would have no effect on the magnitudes of the three roots of this equation.

PROBLEMS

1-1 (a) Using the stress cubic equation, determine the *magnitudes* of the three principal stresses for the following stress state where all stresses are in ksi.

$$\begin{vmatrix} 2 & \frac{1}{4} & \frac{1}{5} \\ \frac{1}{4} & 1 & 0.65 \\ \frac{1}{5} & 0.65 & -\frac{1}{2} \end{vmatrix}$$

(b) Check the answers from (a) by use of the stress invariants I_1 and I_2.

(c) Determine the direction cosines for the largest principal stress.

1-2 A cube of metal is subjected to the following stress system, in ksi:

$$\begin{vmatrix} -20 & 0 & -9 \\ 0 & 45 & 0 \\ -9 & 0 & 12 \end{vmatrix}$$

Calculate the magnitude of the largest shear stress that exists in this cube.

1-3 Given the following stress system at a point, in units of MPa:

$$\begin{vmatrix} 100 & 0 & -15 \\ 0 & -55 & 0 \\ -15 & 0 & -20 \end{vmatrix}$$

Calculate:

(a) The magnitude of the maximum principal stress.

(*b*) The direction cosines of the maximum principal stress.

(*c*) Consider a different coordinate system *a*, *b*, *c* at this point, where σ_a = +15 MPa, σ_c = −55 MPa. Calculate σ_b.

1-4 A round bar of $\frac{1}{2}$ in. diameter is subjected to an axial tensile pull of 5000 pounds. A second identical bar is subjected to a fluid pressure around its periphery of 3000 psi then to the same axial load as bar number one. Which bar experiences the largest shear stress? Determine its magnitude.

1-5 Consider the stress states

$$\begin{vmatrix} 10 & 1 & 0 \\ 1 & 5 & 0 \\ 0 & 0 & 4 \end{vmatrix} \quad \text{and} \quad \begin{vmatrix} 6 & 1 & 0 \\ 1 & 1 & 0 \\ 0 & 0 & 0 \end{vmatrix}$$

(*a*) Show that the principal stresses can be determined for both states using Mohr's circle and indicate the magnitudes of $\sigma_1, \sigma_2, \sigma_3$.

(*b*) Compare these two stress states by discussing the similarities between the sets of Mohr's circles.

1-6 The following stress system, in psi, exists at a point:

$$\begin{vmatrix} 6920 & 864 & 691 \\ 864 & 3460 & 2250 \\ 691 & 2250 & -1730 \end{vmatrix}$$

One principal stress is 7300 psi, with direction cosines $k = 0.9433, l = 0.2979,$ $m = 0.1462$. Another principal stress is −2582 psi, with direction cosines $k = 0.0369, l = 0.3441, m = -0.9382$.

(*a*) Find the magnitude of the third principal stress in psi.

(*b*) Sketch and label the three principal stresses in their approximate directions on a coordinate system; and

(*c*) Label (with its direction cosine) at least one of the angles between the 7300 psi stress and any *x*, *y*, or *z* axis, and between the −2582 psi stress and any *x*, *y*, or *z* axis.

1-7 In the fastening of a bolted joint ($\frac{3}{8}$ in. diameter can be used for *all* calculations), a tensile load of 2200 pounds is induced in the bolt due to tightening. Simultaneously, the bolt is subjected to a torque of 165 lb-in. due to the tightening of the nut.

(*a*) Determine the magnitude of the maximum shear stress and indicate the relation of the plane on which it acts with respect to the axis of the bolt.

(*b*) Determine the magnitudes of $\sigma_1, \sigma_2, \sigma_3$.

(*c*) If the torque were not present, by what percent would σ_1 be reduced as compared to that value in part (b)?

1-8 Stresses at a point, in MN/m², are

$$\begin{vmatrix} 50 & 0 & -27 \\ 0 & 60 & 0 \\ -27 & 0 & 80 \end{vmatrix}$$

(*a*) Calculate the maximum normal stress at this point.

(*b*) Draw this maximum normal stress in a sketch and show and calculate the angles it makes with the $+x$, $+y$, and $+z$ axes.

1-9 Determine the magnitudes of the three principal stresses in a thin-walled tube which is subjected simultaneously to a tensile axial stress of 8000 psi and a shear stress due to twisting of 12,000 psi.

1-10 A solid shaft of 2 in. diameter is subjected to an axial tensile load of 10 tons and a torque of 5000 inch-pounds. Find the magnitude of the principal stresses and the single largest shear stress that results.

1-11 Using typical symbolization, show that for moment equilibrium $\tau_{xy} = \tau_{yx}$, etc.

1-12 Consider a stress state where $\sigma_x = 25$ ksi, $\sigma_y = 10$ ksi and $\tau_{xy} = +5$ ksi, using the Mohr circle convention for the signs of the stresses.

(*a*) Determine the magnitudes of the principal stresses both analytically and with a Mohr circle construction.

(*b*) Draw a stress element in the *x-y* plane that shows the principal stresses and the angular rotation of the orientation of this stress element with respect to the *x* axis.

(*c*) Determine the value of the largest shear stress.

1-13 Consider the stress state:

$$\begin{vmatrix} 10 & 1 & 2 \\ 1 & 5 & 3 \\ 2 & 3 & 4 \end{vmatrix}$$

Demonstrate graphically why this cannot be solved for the principal stresses using a Mohr circle approach. This should be done by "looking" down the *x*, *y*, and *z* axes respectively and plotting a Mohr circle that represents the *y-z*, *x-z*, and *x-y* planes respectively.

1-14 Show that if

$$\tau_{max} = \tfrac{1}{2}[(\sigma_x - \sigma_y)^2 + 4\tau_{xy}^2]^{1/2}$$

then:

$$\tau_{max} = \tfrac{1}{2}(\sigma_1 - \sigma_2)$$

1-15 Assuming that yielding is due to a shear mechanism show that if $\sigma_1 = \sigma_2 = \sigma_3 \neq 0$ that yielding should not occur.

1-16 The following shear stresses are said to prevail for a given state of stress:

$$\tau_{13} = 4 \text{ ksi}, \qquad \tau_{12} = 3 \text{ ksi}, \qquad \tau_{23} = 1.5 \text{ ksi}$$

You are asked to specify any combination of principal stresses, $\sigma_1, \sigma_2, \sigma_3$ that could produce those shear stresses. Explain your answer.

1-17 Two pieces are to be glued along a line *ab* as shown below; this joint can be positioned at any angle bounded by $0 < \theta < 60°$. The allowable shear stress of the joint (τ_a) is $\tfrac{3}{4}$ the allowable normal stress acting on the joint (σ_a). What value of θ will permit the largest load, *P*, to be supported? If the permissible variation of θ were from 0 to 45 degrees, how would this influence your answer?

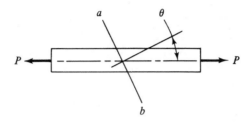

1-18 A bracket is bonded to another surface as shown in the sketch; at the *interface* a maximum shear stress of 5 ksi and maximum tensile stress of 7 ksi can be tolerated. Compression normal to the interface need be of no concern, while the boundary *BC* is stress-free. If *AB* equals 1 in., *BC* equals one and one-half in. and the 1, 2, 3 coordinate system is taken as shown, what is the maximum tensile stress, σ_2, that can be applied before *either* limiting interface stress is reached? Assume that σ_2 causes a homogeneous state of stress in *ABC*.

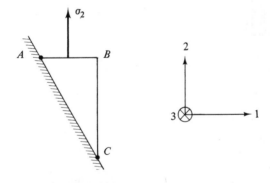

1-19 A stress state is described by

$$\begin{vmatrix} 20 & -6 & 10 \\ -6 & 10 & 8 \\ 10 & 8 & 7 \end{vmatrix}$$

all units in MPa.

(a) Determine the principal stress values.

(b) If the shear stress shown as +10 were −10 instead and all other values are as shown, find the principal stresses.

2

STRAIN

2-1 INTRODUCTION

When a body is deformed by external forces, points in the body may be displaced from the positions they occupied under no load. Strain is related to such displacements. Pure rigid-body motions, such as translation or rotation, do not cause a body to be strained. Consider Fig. 2-1 where the initial length of AB, denoted as l_0, is displaced to $A'B'$ (l') due to external forces. Three possibilities exist:

1. l_0 is equal in length and parallel to l'; thus *translation* occurred but there is no strain induced.
2. l_0 is equal in length to l' but they are no longer parallel; thus, *rotation* occurred but no strain was induced.
3. l_0 and l' are unequal in length so a state of strain has been induced.

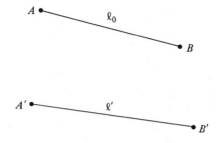

Figure 2-1 Displacement of an arbitrary length l_0 to length l'.

The displacements of A to A' and B to B' are used to define the magnitude of strain and it should be noted that both translation and rotation may have occurred during the displacement.

When condition three has been satisfied, the strain in AB is *defined* as:

$$e \equiv \frac{l' - l_0}{l_0} = \frac{\Delta l}{l_0} \qquad (2\text{-}1)$$

and by choosing small values of these lengths, the concept or mathematical notion of the *state of strain at a point* may be invoked in a manner similar to that used with stress at a point in Chap. 1. Tensile strains, caused by extensions, are considered positive whereas compressive strains, due to shortening, are negative. Shear strains, due to angular distortion, are defined as negative or positive in the next section.

For reasonable accuracy, the definition of strain as in Eq. (2-1) is limited to small, infinitesimal, or incremental amounts of displacement. When large displacements are encountered, lines that were initially straight may become curved due to internal rotation. In such cases, the usual strain equations become either so complicated or highly suspect in terms of accurate description of the situation, that they find little practical use. In essence, during large-scale straining it is usually impossible to document the actual *strain history* that a small segment has undergone. Without such knowledge, an accurate calculation of the induced strain is impossible. On the other hand, if a large deformation occurs along a known strain path, the induced strain can be determined, especially if the path is linear (i.e., not accompanied by internal rotation) and shows no *reversal* of direction.

2-2 THE TWO-DIMENSIONAL CASE FOR SMALL STRAINS

In the following developments it is assumed that displacements are small, initially straight lines remain straight and any higher order terms are negligible. For simplicity, consider a two-dimensional case as illustrated in Fig. 2-2.

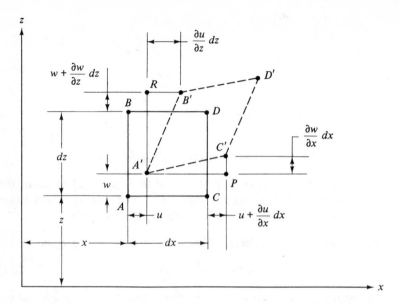

Figure 2-2 Description of the case of small strains in two dimensions.

Initially the element of concern is ABCD where $A(x, z)$, $C(x + dx, z)$, $B(x, z + dz)$ and $D(x + dx, z + dz)$ define the corners of the element. Due to external effects the element distorts to $A'B'C'D'$ where:

$$A'(x + u, z + w), \qquad C'\left(x + dx + u + \frac{\partial u}{\partial x}dx, z + w + \frac{\partial w}{\partial x}dx\right)$$

define the ends of the line whose initial length was AC.

Physically, *small* displacements means $A'P \approx A'C'$ and $\sphericalangle\, PA'C'$ is also *small* (i.e., $\tan \theta \approx \theta$).

$$e_{xx} \equiv \frac{A'C'}{AC} - 1 \approx \frac{A'P}{AC} - 1 = \frac{dx - u + u + \frac{\partial u}{\partial x}dx}{dx} - 1$$

so,

$$e_{xx} = 1 + \frac{\partial u}{\partial x} - 1 = \frac{\partial u}{\partial x}$$

and similarly

$$e_{zz} = \frac{\partial w}{\partial z}$$

Now for shear strain associated with A, this is defined as the change in the angle CAB. For now, this shear strain is denoted as γ_{xz} since it occurs in the x-z plane and by convention, if this angle increases γ is negative whereas a decrease in this angle implies that γ is positive.

The total change in angle CAB is $RA'B' + PA'C'$. Recalling that $\tan \theta \approx \theta$ with our assumptions,

$$\text{Angle } PA'C' = \text{arc tan } \frac{\dfrac{\partial w}{\partial x} dx}{A'P} = \text{arc tan } \frac{\dfrac{\partial w}{\partial x} dx}{dx + \dfrac{\partial u}{\partial x} dx}$$

so

$$\text{Angle } PA'C' = \frac{\dfrac{\partial w}{\partial x}}{1 + \dfrac{\partial u}{\partial x}} = \frac{\partial w}{\partial x} \quad \text{since } \frac{\partial u}{\partial x} \ll 1$$

Similarly, angle $RA'B' = \partial u/\partial z$ so $\gamma_{xz} = \partial w/\partial x + \partial u/\partial z$, where u and w are displacements in the x and z directions.

The full expressions for the strain components at a point could be defined by the *strain tensor*, that is,

$$e_{ij} = \begin{vmatrix} e_{xx} & e_{yx} & e_{zx} \\ e_{xy} & e_{yy} & e_{zy} \\ e_{xz} & e_{yz} & e_{zz} \end{vmatrix} \tag{2-2}$$

Here however, $e_{xz} \equiv \frac{1}{2}\gamma_{xz} = \frac{1}{2}(\partial w/\partial x + \partial u/\partial z)$. The γ form is the shear strain associated with an applied shear stress and the shear modulus; it is usually referred to as *engineering shear strain*. To cast the strain tensor it is necessary to *define* a *mathematical* shear strain equal to half γ. Then, the form described in Eq. (2-2) permits tensor transformation rules to be used.

For the purposes that follow, the symbolical form of strains to be employed is:

1. Normal strains, whether tensile or compressive, are described as $e_x, e_y,$ and e_z.
2. Shear strains are described by $\gamma_{xy}, \gamma_{yz},$ and γ_{zx} keeping in mind that $\frac{1}{2}\gamma$ must be used when transformations are involved. Several points and observations are worth noting.
   a. With an interchange of subscripts it should be apparent that in the general case, the normal strains are expressed by the displacements u, v, and w in the x, y, and z directions as follows:

$$e_x = \frac{\partial u}{\partial x}, \qquad e_y = \frac{\partial v}{\partial y}, \qquad \text{and} \qquad e_z = \frac{\partial w}{\partial z} \tag{2-3}$$

   b. For shear strains,

$$\gamma_{xy} = \frac{\partial u}{\partial y} + \frac{\partial v}{\partial x}, \qquad \gamma_{yz} = \frac{\partial w}{\partial y} + \frac{\partial v}{\partial z},$$

   and

$$\gamma_{xz} = \frac{\partial w}{\partial x} + \frac{\partial u}{\partial z} \tag{2-4}$$

c. Since the strains defined in Eqs. (2-3) and (2-4) are functions
of the displacements u, v, and w they are not independent of
one another but must be interrelated through *compatibility*
equations. Because no use will be made of these equations
(since they are really of importance in a theory-of-elasticity
approach) they will not be presented here. Ford [1] may be
consulted for details. A physical interpretation of compatibility
is, however, of importance. It signifies that during displacements, points adjacent to each other stay adjacent. In regard
to Fig. 2-1, all points between A and B end up in a corresponding alignment between A' and B' such that no "holes"
would result.

2-3   THE THREE-DIMENSIONAL CASE FOR SMALL STRAINS

Here the procedure in Ford [1] is followed and the purpose of this section
is both for completeness' sake and a better understanding of why the usual
strain equations are limited in their use. Figure 2-3 shows the three-dimensional model used to define the strain at a point and the concern here is to
determine the strain induced in AC after it ends up at $A'C'$.

The strain in $AC \equiv (A'C'/AC) - 1$ (this assumes all particles on AC
move to $A'C'$).

1. Draw $A'K$ through A' and parallel to AC and XOY. The angle
$KA'C'$ is the rotation of AC to $A'C'$ which produces components
in both the XOY and XOZ planes.

2. Project $A'C'$ and $A'K$ on the XOY plane as EG and EF. Thus,
$A'K = EF = dx + (\partial u/\partial x)\, dx$. However, $EG > EF$ because the
displacement v may change as a function of x. This indicates
$FG = (\partial v/\partial x)\, dx = KL$.

3. In a similar manner, displacement w may be a function of x such
that $C'L = (\partial w/\partial x)\, dx$. Thus,

$$(A'C')^2 = (EF)^2 + (FG)^2 + (C'L)^2$$

or

$$(A'C')^2 = \left(dx + \frac{\partial u}{\partial x}\, dx\right)^2 + \left(\frac{\partial v}{\partial x}\, dx\right)^2 + \left(\frac{\partial w}{\partial x}\, dx\right)^2$$

or

$$(A'C')^2 = (dx)^2\left[1 + 2\frac{\partial u}{\partial x} + \left(\frac{\partial u}{\partial x}\right)^2 + \left(\frac{\partial v}{\partial x}\right)^2 + \left(\frac{\partial w}{\partial x}\right)^2\right]$$

Now *define*

$$e_{xx} = \frac{\partial u}{\partial x} + \frac{1}{2}\left[\left(\frac{\partial u}{\partial x}\right)^2 + \left(\frac{\partial v}{\partial x}\right)^2 + \left(\frac{\partial w}{\partial x}\right)^2\right]$$

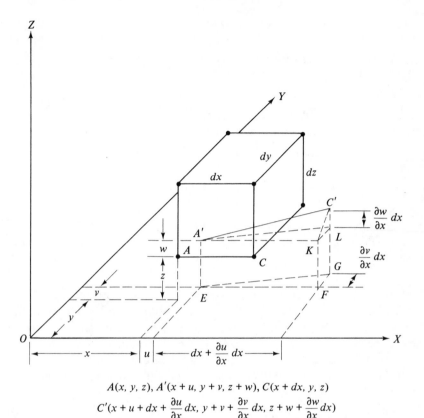

$$A(x, y, z), A'(x + u, y + v, z + w), C(x + dx, y, z)$$

$$C'(x + u + dx + \frac{\partial u}{\partial x} dx, y + v + \frac{\partial v}{\partial x} dx, z + w + \frac{\partial w}{\partial x} dx)$$

Figure 2-3 Small strains for the three-dimensional case.

thus, the strain in $AC = (A'C'/AC) - 1 = (dx \sqrt{1 + 2e_{xx}}/dx) - 1$ or, the strain in $AC = (1 + 2e_{xx})^{1/2} - 1$. Expand $(1 + 2e_{xx})^{1/2}$ to give the strain in $AC = 1 + e_{xx} +$ higher orders $- 1 \approx e_{xx}$. Finally, the strain in

$$AC \approx e_{xx} = \frac{\partial u}{\partial x} + \frac{1}{2}\left\{ \left(\frac{\partial u}{\partial x}\right)^2 + \cdots \right\}$$

and for *small* displacements $(\partial u/\partial x) \gg (\partial u/\partial x)^2$ etc., so that

$$e_{xx} = \frac{\partial u}{\partial x}, \qquad e_{yy} = \frac{\partial v}{\partial y}, \qquad e_{zz} = \frac{\partial w}{\partial z} \qquad (2\text{-}5)$$

as shown earlier. The entire point of this exercise is to indicate what is ignored or discarded when Eq. (2-5) is used. For large rotations, the higher order terms may not be negligible so the accuracy of Eq. (2-5) is questionable. Again it must be realized however, that only for a limited number of problems have the full strain equations found practical use!

2-4 STRAIN TRANSFORMATIONS

Transformation of strains from one set of reference axes to another set follows a development identical to that for stresses, so a repetition of that development is unessential.

The resulting expressions take the following forms:

$$e_{x'x'} = l^2_{x'x}e_{xx} + l^2_{x'y}e_{yy} + l^2_{x'z}e_{zz} + l_{x'y}l_{x'z}e_{yz} + l_{x'z}l_{x'x}e_{zx}$$
$$+ l_{x'x}l_{x'y}e_{xy} + l_{x'z}l_{x'y}e_{zy} + l_{x'x}l_{x'z}e_{xz} + l_{x'y}l_{x'x}e_{yx} \quad (2\text{-}6)$$

and

$$e_{y'z'} = l_{y'x}l_{z'x}e_{xx} + l_{y'y}l_{z'y}e_{yy} + l_{y'z}l_{z'z}e_{zz} + l_{y'y}l_{z'z}e_{yz} + l_{y'z}l_{z'x}e_{zx}$$
$$+ l_{y'x}l_{z'y}e_{xy} + l_{y'z}l_{z'y}e_{zy} + l_{y'x}l_{z'z}e_{xz} + l_{y'y}l_{z'x}e_{yx} \quad (2\text{-}7)$$

and similarly for $e_{y'y'}$, $e_{x'z'}$ and so forth. In Eqs. (2-6) and (2-7) the *mathematical definition* of shear strain is used since these are tensor transformations where:

$$e_{ij} = l_{im}l_{jn}e_{mn}$$

with the usual meaning. In this form, strains may be plotted using a Mohr circle approach in a manner identical to that described earlier with regard to stresses and since the circle plot is a *graphical tensor transformation*, shear strains equal to $\frac{1}{2}\gamma$ must be plotted along the ordinate. Thus $2e_{ij} = \gamma_{ij}$ where γ_{ij} is the shear strain common in engineering. The equivalent transformations in terms of γ (and noting that $e_{ij} = e_{ji}$ and $\gamma_{ij} = \gamma_{ji}$) become:

$$e_{x'x'} = l^2_{x'x}e_{xx} + l^2_{x'y}e_{yy} + l^2_{x'z}e_{zz} + l_{x'y}l_{x'z}\gamma_{yz}$$
$$+ l_{x'z}l_{x'x}\gamma_{zx} + l_{x'x}l_{x'y}\gamma_{xy} \quad (2\text{-}8)$$

and

$$\gamma_{y'z'} = 2(l_{y'x}l_{z'x}e_{xx} + l_{y'y}l_{z'y}e_{yy} + l_{y'z}l_{z'z}e_{zz}) + \gamma_{yz}(l_{y'y}l_{z'z} + l_{y'z}l_{z'y})$$
$$+ \gamma_{zx}(l_{y'z}l_{z'x} + l_{y'x}l_{z'z}) + \gamma_{xy}(l_{y'x}l_{z'y} + l_{y'y}l_{z'x}) \quad (2\text{-}9)$$

and similarly for $e_{y'y'}$, $\gamma_{x'z'}$ and so forth.

It must be understood that Eqs. (2-8) and (2-9) are not tensor transformations. Rather, they make use of the usual measured value of shear strain, γ, instead of the defined form of $\frac{1}{2}\gamma$.

As with stresses, there exists a set of axes along which the shear strains are zero. These are the principal strain axes and these normal strains are the principal strains.

Where one axis is common to both sets of axes (i.e., z and z' coincide), strain transformations take the form:

$$e_{xx} = e_{x'x'}\cos^2\theta + e_{y'y'}\sin^2\theta + \gamma_{x'y'}\cos\theta\sin\theta \quad (2\text{-}10)$$

$$e_{yy} = e_{x'x'}\sin^2\theta + e_{y'y'}\cos^2\theta - \gamma_{x'y'}\cos\theta\sin\theta \quad (2\text{-}11)$$

$$\gamma_{xy} = 2e_{x'x'}\cos\theta\sin\theta - 2e_{y'y'}\cos\theta\sin\theta - \gamma_{x'y'}(\cos^2\theta - \sin^2\theta) \quad (2\text{-}12)$$

38

By using appropriate trigonometric identities and recalling that $\gamma_{xy} = 2e_{xy}$ by definition, the similarity of Eq. (2-10) with (1-13) and (2-12) with (1-15) can be shown.

In terms of principal strains several useful relations are:

$$e_{xx} = e_x = \frac{e_1 + e_2}{2} + \frac{e_1 - e_2}{2} \cos 2\theta \qquad (2\text{-}13)$$

$$e_{yy} = e_y = \frac{e_1 + e_2}{2} - \frac{e_1 - e_2}{2} \cos 2\theta \qquad (2\text{-}14)$$

$$\gamma_{xy} = (e_1 - e_2) \sin 2\theta \qquad (2\text{-}15)$$

Example 2-1.

The sketch (Fig. E2-1) shows three wire resistance strain gages attached to the surface of a part. Under loads, the gages give the following indications of strains:

$$\text{Gage (a)} = 0.002 \text{ in./in.}$$
$$\text{Gage (b)} = 0.0025 \text{ in./in.}$$
$$\text{Gage (c)} = 0.0005 \text{ in./in.}$$

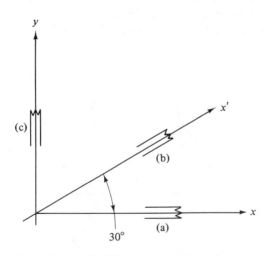

Find the magnitude of γ_{xy} and of the principal strains in the x-y plane.

Solution.

Using Eq. (2-8), where $l_{x'x} = \cos 30°$ and $l_{x'y} = \cos 60°$, $0.0025 = 0.002(0.866)^2 + (0.0005)(0.5)^2 + \gamma_{xy}(0.866)(0.5)$, so, $\gamma_{xy} = 0.00202$ in./in.

To find the principal strains, Eqs. (2-8) and (2-9) can be rearranged in a manner that led to Eq. (1-18) to give:

$$e_{1,2} = \frac{e_x + e_y}{2} \pm \left[\left(\frac{e_x - e_y}{2}\right)^2 + \left(\frac{\gamma_{xy}}{2}\right)^2\right]^{1/2}$$

$$e_{1,2} = \left(\frac{0.002 + 0.0005}{2}\right) \pm \left[\left(\frac{0.002 - 0.0005}{2}\right)^2 + \left(\frac{0.00202}{2}\right)^2\right]^{1/2}$$

$$e_1 = 0.00251 \text{ in./in.,} \qquad e_2 = 0$$

2-5 MOHR'S CIRCLE FOR STRAINS

By using the mathematical definition of shear strain it was mentioned earlier that a tensor form of strain results and is, in fact, identical in concept to that used for stresses. It is, therefore, unnecessary to repeat the earlier development and detail of the section devoted to Mohr's circle. Instead, some useful relationships are noted and may be compared with their stresses except that $\frac{1}{2}\gamma$ must be plotted along the ordinate. Note the following:

1. The center of the circle is located at $\frac{1}{2}(e_1 + e_2)$ or $\frac{1}{2}(e_x + e_y)$.
2. The radius of the circle is equal to $\frac{1}{2}(e_1 - e_2)$
3. $(e_1 - e_2) = [(e_x - e_y)^2 + \gamma_{xy}^2]^{1/2}$

A number of other identities, developed directly from those involving stresses, could be presented. The reader may find it of interest to do so.

Figure 2-4 presents two illustrations of the Mohr's circle plots for stan-

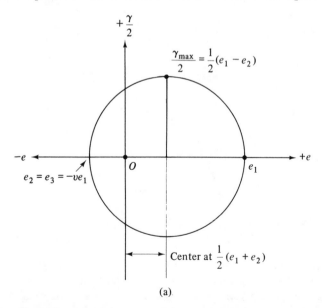

(a)

Figure 2-4 Mohr's circle for strains due to (a) uniaxial tension and (b) a generalized case of loading.

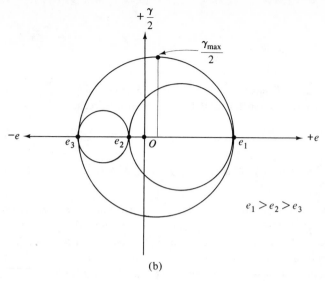

(b)

Fig. 2-4 (cont.)

dard uniaxial tension and where three different but nonzero principal strains occur (v is Poisson's ratio).

Example 2-2.

With the values of e_x and e_y given in Example 2-1 and using the computed value of $\gamma_{xy} = 0.00202$, find the principal strains in the x-y plane using a Mohr's circle construction.

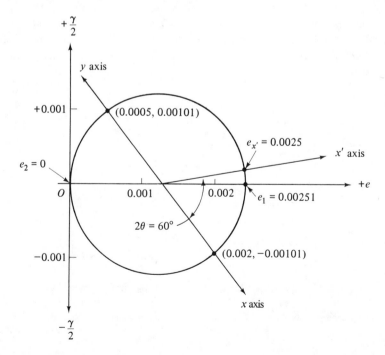

Solution.

Noting that γ_{xy} is positive by the convention used in regard to the stress tensor, it must act counterclockwise and so is negative in regard to a plot of Mohr's circle.

2-6 ENGINEERING STRAIN AND TRUE STRAIN

Up to this point, it has been assumed that deformations and resulting strains are small; elastic strains satisfy this requirement. However, under large plastic strain, this condition is no longer satisfied and a more realistic definition of strain is utilized. Several terms are used to discuss small strains, two of them being *engineering* or *nominal*. Throughout this text, such strains will carry the symbol e and are defined as:

$$e = \frac{\partial u}{\partial x} = \frac{\Delta l}{l_0} = \frac{l - l_0}{l_0} = \frac{l}{l_0} - 1 \qquad (2\text{-}16)^*$$

To show the limitation of the above definition, consider a bar of length l_0 that is loaded in tension until its length is doubled. This would induce an extension of 100 percent and a strain of unity. If an identical bar were to be compressed 100 percent, it would have to be flattened to zero thickness! Obviously the latter operation would have altered the structure to a far more severe degree yet this compressive strain would be unity also. Certainly the real strain is not equivalent in both cases even though the *magnitudes* are equivalent. To avoid this, the concept of *true* strain (also called *logarithmic* or *natural*) is used. The basis of this is given in Fig. 2-5 and the accompanying discussion.

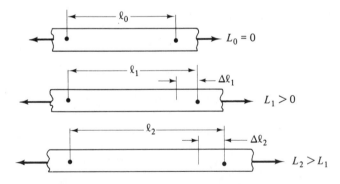

Figure 2-5 Basis for defining true or logarithmic strain.

*Where e represents the base of natural logarithms, it will be so indicated.

True strain, as first proposed in 1909 by Ludwik, [2] is interpreted as an incremental elongation divided by the gage length that existed just prior to this increment of change. Consider the sequence of events depicted in Fig. 2-5.

$$\text{True strain} = \epsilon \equiv \frac{\text{incremental length increase}}{\text{previous gage length}} \qquad (2\text{-}17)$$

As the load is increased from zero to L_1, the true strain that exists due to L_1 would be:

$$\epsilon_1 = \frac{\Delta l_1}{l_0}$$

Now as the load is increased from L_1 to L_2, the added (incremental) true strain induced would be:

$$\epsilon_2 = \frac{\Delta l_2}{l_0 + \Delta l_1} = \frac{\Delta l_2}{l_1}$$

Since the denominator reflects the previous change in length caused by the initial strain increment, it is obvious why true strain differs from nominal strain. Now, the total true strain induced by load L_2 can be thought of as being the sum of ϵ_1 and ϵ_2, so

$$\epsilon_t = \epsilon_1 + \epsilon_2 = \frac{\Delta l_1}{l_0} + \frac{\Delta l_2}{l_0 + \Delta l_1}$$

or in a general form:

$$\epsilon_t = \frac{\Delta l_1}{l_0} + \frac{\Delta l_2}{l_0 + \Delta l_1} + \cdots$$

up to the particular load of concern. Thus, dropping the subscript t for total:

$$\epsilon = \Sigma \frac{\Delta l_i}{l_0 + \Delta l_{i-1}} \qquad (2\text{-}18)$$

which can be readily expressed as

$$d\epsilon \equiv \frac{dl}{l} \qquad (2\text{-}19)$$

Assuming the specimen contains no initial strain, there results:

$$\int_0^{\epsilon_t} d\epsilon = \int_{l_0}^{l_t} \frac{dl}{l} \quad \text{or} \quad \epsilon = \ln\left(\frac{l}{l_0}\right) \qquad (2\text{-}20)$$

It should be noted that by this definition true strains are additive so long as each incremental strain increase is induced *along the same path* as all previous increments. This will receive greater attention in Chap. 5.

Return to the discussion below Eq. (2-16). Now if l_0 is doubled,

$$\epsilon = \ln\left(\tfrac{2}{1}\right) = +0.693$$

To induce the same *magnitude* of strain in compression,

$$\epsilon = \ln\left(\frac{0.5}{1}\right) = -\ln 2 = -0.693$$

Thus, the length l_0 is compressed to only one-half its initial value to impart the same true strain in compression as resulted in tension when the length was doubled. This makes much greater physical sense.* Two final points are worth noting:

1. The symbol ϵ or its incremental form $d\epsilon$ will *always* be used when plastic strains are involved whereas e will be used when elastic strains are discussed.
2. The tensor concept is valid for small deformations so that incremental true strains $(d\epsilon)$ are tensor quantities regardless of the overall degree of deformation.

Example 2-3.

The initial length of a workpiece is 50 mm when the part is under no load. Under the application of a tensile load, the length is increased to 60 mm. Find the nominal and true strains induced.

Solution.

$$\text{Nominal strain, } e = \frac{60 - 50}{50} = 0.200$$

$$\text{True strain, } \epsilon = \ell n\left(\frac{60}{50}\right) = 0.182$$

Example 2-4.

If the loading in Example 2-3 had been compressive instead of tensile, to what length would the 50 mm dimension be reduced to produce the equivalent true strain? What would be the nominal strain under this condition?

Solution.

$$\epsilon = -0.182 = \ell n\left(\frac{l}{50}\right), \qquad +0.182 = \ell n\left(\frac{50}{l}\right)$$

so

$$e^{0.182} = \frac{50}{l} \qquad \text{where } e \text{ is the base for natural logarithms.}$$

$$1.2 = \frac{50}{l}, \qquad l = 41.67 \text{ mm}$$

As for the nominal strain,

$$e = \frac{l - l_0}{l_0} = \frac{41.67 - 50}{50} = -0.167 \qquad \text{(here, } e \text{ is the nominal strain)}$$

The key point here is that for strains well outside the elastic region, equivalence in the magnitudes of true strains will not result in equivalence of nominal strains (or vice versa).

*As discussed in Chap. 4, the volume of material remains essentially constant during plastic deformation. The use of true strains satisfies this observation.

REFERENCES

1. H. FORD, *Advanced Mechanics of Materials* (New York: John Wiley and Sons, Inc., 1963), pp. 129–31, 115–21.

2. P. LUDWIK, *Elemente der technologischen Mechanik* (Berlin: Springer, 1909).

PROBLEMS

2-1 Show that the expression for e_x given in Eq. (2-13) is correct.

2-2 A block of initial dimensions l_0, w_0, t_0 is subjected to tensile loading that increases l_0 to l while the other dimensions decrease to w and t respectively. Using the definition of true strain given in Eq. (2-19), show that constancy of volume requires that

$$d\epsilon_1 + d\epsilon_2 + d\epsilon_3 = 0.$$

2-3 In Example 2-1 use was made of the expression

$$e_{1,2} = \tfrac{1}{2}(e_x + e_y) \pm \tfrac{1}{2}[(e_x - e_y)^2 + \gamma_{xy}^2]^{1/2}$$

Show how this relationship can be derived.

2-4 Suppose the block in Prob. 2-2 deformed under constant volume and the length, l_0, were doubled.
   (a) Calculate the engineering strains in the other directions.
   (b) Calculate the true strains in the three directions.
   (c) Comment on the difference in volume changes based upon these two methods of determining strains.

2-5 Wire is produced by pulling it through a series of circular dies of ever-decreasing cross-sectional area until the desired wire diameter is reached. If a wire of initial diameter of 0.010 in. is pasesd through five dies, each causing an *area* reduction of 10 percent (this is called drawing),
   (a) What is the final wire diameter?
   (b) Plot the ratio of d/d_0 versus the total nominal *and* true strains resulting after each die has been passed.

2-6 Three strain gages are positioned on a solid such that gage "1" lies parallel to the "x axis" (this is an arbitrary designation). Gage 2 is positioned 60° counterclockwise to gage one while the third gage is positioned 120° counterclockwise to the first. Under loading, the gages indicate the following strains,

$$\text{Gage 1} = 3000 \ \mu \ \text{inch/inch}$$

$$\text{Gage 2} = 1500 \ \mu \ \text{inch/inch}$$

$$\text{Gage 3} = 1000 \ \mu \ \text{inch/inch}$$

With the use of a Mohr circle plot, determine the principal strains in the plane from which these measured values were obtained.

2-7 A solid is deformed under "plane strain" conditions (i.e., $e_2 = 0$). Measurements indicate that strains in the 1–3 plane (perpendicular to the "2" direction) are:

$$e_x = 0.010, \qquad e_y = 0.005, \qquad \gamma_{xy} = 0.007$$

(*a*) Construct Mohr's circle for this condition.

(*b*) Determine the magnitudes of e_1 and e_3.

2-8 Show that true strain, defined as $\ell n \, (l/l_0)$ may be represented by any of the following forms for uniform deformation:

$$\epsilon = \ell n \left(\frac{l}{l_0}\right) = \ell n \left(\frac{A_0}{A}\right) = 2 \, \ell n \left(\frac{D_0}{D}\right) = \ell n \left(\frac{1}{1-r}\right)$$

where l_0, A_0, and D_0 are original values of length, area and diameter, l, A, and D are instantaneous values, and r is the decimal reduction in area.

3

ISOTROPIC ELASTICITY

3-1 INTRODUCTION

In the two previous chapters it was suggested that external loads cause a solid to be stressed and strained yet the equations developed in regard to stresses contained no strains and vice-versa. Intuitively it should be expected that a relationship between stress and strain should exist and experience has shown this to be true. The coupling of stress and strain, leading to what are usually called *constitutive relations*, has been found to be dependent upon particular properties of the solid in question. When these properties are found to be the same regardless of the direction of measurement, the solid is said to be *isotropic*. If, however, the value of a property differs as a function of direction, the solid is classed as *anisotropic* with regard to the property of concern. A solid is said to be *homogeneous* if its properties do not vary with location within the body. Since the emphasis of this text is on the mechanical behavior of solids, only mechanical properties such as elastic modulus, Poisson's ratio, strength, etc., are of concern.

The thrust of this chapter is concerned with the constitutive relationships for the elastic behavior of a solid that is both isotropic and homogeneous. Although the assumptions invoked may not find exact agreement with the behavior of real solids, the differences involved are often so minor that inaccuracies become minimal.

3-2   MODEL OF A PERFECTLY ELASTIC SOLID

To provide a physical feel for the mechanical behavior of solids, it is useful to begin with a simple model. It must be realized, of course, that the actual internal structure of solids is far more complex than this model might imply; in addition, the model finds meaning only to the extent that it describes the observed macroscopic behavior of the material in question.

For a perfectly elastic solid, we consider the model shown in Fig. 3-1; it involves a simple spring. Certain assumptions and comments are pertinent to Fig. 3-1:

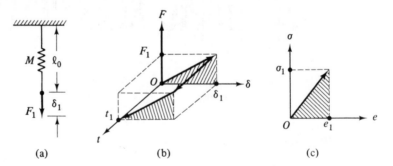

(a)   (b)   (c)

Figure 3-1 Description of a perfectly elastic solid showing the model (a), the force-displacement-time plot (b), and the stress-strain plot (c).

1. Prior to the application of the force F_1, the solid has a length l_0 and a property whose resistance tó extension under load is defined by M. This property is analogous to the modulus of elasticity and does not vary with time.

2. Under load F_1, the body extends an amount δ_1 and this extension takes place instantaneously and linearly.

3. The energy stored in the body is shown as the shaded area in the F-δ plane.

4. As F_1 is applied for the time t_1, δ_1 remains constant.

5. Upon release of F_1 instantaneous recovery occurs and all of the stored energy is released.

6. The same analysis would hold if l_0 were shortened rather than extended.

7. The equivalent σ-e curve is shown in Fig. 3-1(c).

Since stress and strain provide more meaningful parameters than force and extension (or contraction), an equivalent stress-strain vs. time plot could be constructed for this situation. The slope of the σ-e plot is the modulus of elasticity E which is a measure of the stiffness or rigidity of a solid. As to the shaded area under the σ-e plot at zero time, this should be viewed as the *strain energy* stored in the body under load.

3-3 CONSTITUTIVE RELATIONS

Consider a solid subjected to a uniaxial force. Under elastic deformation as described by Fig. 3-1, the stress and strain that result are directly related by the elastic modulus. This is known as Hooke's law and in its simplest form is expressed as $\sigma = Ee$. A generalized form of this relationship, extended to three dimensions, is found to have wider use. Experiments show that as the tensile stress σ_1 is applied it is coupled with a tensile strain e_1 and with contractions in two perpendicular directions in the plane at right angles to the line of force application. Note that either an x-y-z or 1-2-3 coordinate system may be introduced. The strains in the 2 and 3 directions are directly related to the strain e_1 through the parameter called Poisson's ratio, v. Thus,

$$e_2 = e_3 = -ve_1$$

since the contractions cause negative strains and the equality of these two results from isotropy.

In a more general situation, tensile stresses might also be applied simultaneously in the three directions, each stress causing a negative strain in the other two directions. Because all equations are linear, the principle of superposition can be invoked and the effects in a given direction simply added together. This means:

$$e_1 = \frac{\sigma_1}{E} - v\frac{\sigma_2}{E} - v\frac{\sigma_3}{E} = \frac{1}{E}[\sigma_1 - v(\sigma_2 + \sigma_3)] \tag{3-1}$$

Note that in uniaxial tension, $\sigma_2 = \sigma_3 = 0$, so Eq. (3-1) degenerates to the usual form of Hooke's law. Identical expressions for e_2 and e_3 result by an appropriate change of subscripts. If an x-y-z system is employed, proper subscript interchange can be introduced into Eq. (3-1) thus:

$$e_x = \frac{1}{E}[\sigma_x - v(\sigma_y + \sigma_z)] \quad \text{and so forth.} \tag{3-2}$$

Regarding shear stress and strain, these are related by

$$\tau_{xy} = G\gamma_{xy} = 2Ge_{xy} \qquad (3\text{-}3)$$

where G is the shear modulus and the *tensor* shear strain is one-half the *engineering* shear strain as discussed in Chap. 2.

Example 3-1.

A material whose elastic modulus is 207 GPa is subjected to three normal stresses σ_1, σ_2, and σ_3 whose magnitudes are 250, 200, and -100 MPa respectively. Determine the normal strains in the three directions. Assume Poisson's ratio is 0.3.

Solution.

Equation (3-1) is appropriate here if all deformation is elastic. The modulus is indicated in MPa as are all stresses:

$$e_1 = \frac{10^{-3}}{207}[250 - 0.3(200 - 100)] = +0.00106$$

$$e_2 = \frac{10^{-3}}{207}[200 - 0.3(250 - 100)] = +0.00075$$

$$e_3 = \frac{10^{-3}}{207}[-100 - 0.3(250 + 200) = -0.00114$$

3-4 ELASTIC CONSTANTS*

Note that shear strains cause only a change in shape while any volume change is produced by normal strains. The latter is related to the *bulk modulus, B*, and the foregoing comments reduce to the following:

1. $E =$ elastic modulus, which relates normal stresses and strains.
2. $G =$ shear modulus, which relates shear stresses and strains.
3. $\nu =$ Poisson's ratio, which relates longitudinal and transverse strains.
4. $B =$ bulk modulus, which relates the mean normal stress (σ_m) to the volume strain or *dilatation*, Δ.

Although four constants are indicated (these should be considered as mechanical properties) only two are independent, that is, knowning any two, the remaining two may be derived as will now be shown.

The fractional volume change or dilatation is defined as follows:

for aunit cubic :

$$\frac{dV}{V} = \Delta \equiv e_1 + e_2 + e_3 = \frac{1 - 2\nu}{E}[\sigma_1 + \sigma_2 + \sigma_3] \qquad (3\text{-}4)$$

*See Polakowski and Ripling [1] for a discussion of the elastic constants for anisotropic materials.

which results using Eq. (3-1). The mean normal stress is:

$$\sigma_m \equiv \frac{1}{3}(\sigma_1 + \sigma_2 + \sigma_3) = \frac{\Delta E}{3(1 - 2v)} \tag{3-5}$$

from use of Eq. (3-4). The bulk modulus is defined as:

$$B \equiv \frac{\sigma_m}{\Delta} = \frac{E}{3(1 - 2v)} \tag{3-6}$$

from use of Eq. (3-5).

Example 3-2.

(a) With the findings in Example 3-1 and Eq. (3-4) find the dilatation.
(b) Repeat part (a) using Eq. (3-5).

Solution.

(a) $\Delta = e_1 + e_2 + e_3 = +0.00181 - 0.00114 = +0.00067$

(b) $\sigma_m = \dfrac{\sigma_1 + \sigma_2 + \sigma_3}{3} = \dfrac{+350}{3} = \dfrac{\Delta E}{3(1 - 2v)}$

so,

$$\Delta = \frac{350(1 - 0.6)}{207 \times 10^3} = +0.00067$$

The positive sign means an increase in volume of the body subjected to these three normal stresses.

If the case of pure shear is considered, the Mohr circles for stress and strain are shown in Fig. 3-2. We see from Fig. 3-2 that:

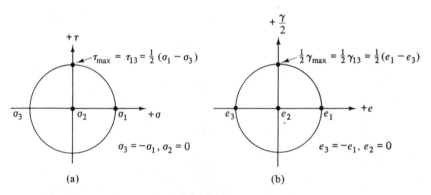

(a)   (b)

Figure 3-2 Mohr's circle of stress (a) and strain (b) for the case of pure shear.

$$\gamma_{13} = e_1 - e_3 = 2e_1$$

From Eq. (3-1):

$$e_1 = \frac{1}{E}[\sigma_1 - v(\sigma_2 + \sigma_3)] = \frac{\sigma_1}{E}(1 + v) \qquad \text{since } \sigma_3 = -\sigma_1, \sigma_2 = 0$$

So,

$$\gamma_{13} = 2e_1 = \frac{2\sigma_1}{E}(1+v) = \frac{\tau_{13}}{G} = \frac{\sigma_1 - \sigma_3}{2G} = \frac{\sigma_1}{G} \qquad \text{since } \sigma_3 = -\sigma_1$$

Therefore,

$$\gamma_{13} = \frac{2\sigma_1}{E}(1+v) = \frac{\sigma_1}{G} \qquad \text{or} \qquad G = \frac{E}{2(1+v)} \qquad (3\text{-}7)$$

With Eqs. (3-6) and (3-7) the interrelationship of the four elastic constants is evident. It is of interest to note that $\tau_{13} = (\sigma_1 - \sigma_3)/2$ whereas $\gamma_{13} = e_1 - e_3$. Again to develop a tensor shear strain, e_{13}, the value of $\frac{1}{2}\gamma_{13}$ must be used. Now, $\tau_{13} = \frac{1}{2}(\sigma_1 - \sigma_3)$ and $e_{13} = \frac{1}{2}(e_1 - e_3)$ so the graphical tensor transformations shown by Fig. 3-2 are identical in form.

3-5 RELATIONSHIP OF MOHR'S CIRCLE FOR STRESS AND STRAIN

It is instructive to show that if the stress circle in Fig. 3-2(a) is known, the corresponding strain circle can be developed. Although in general it is faster to determine the strains using Eq. (3-1), once the stresses and appropriate constants are known, the brief explanation that follows is still worthwhile. In Chap. 4, direct use of this circle correspondence will become most valuable. For the general case,

$$e_1 = \frac{1}{E}[\sigma_1 - v(\sigma_2 + \sigma_3)] \qquad \text{and} \qquad \sigma_m = \frac{1}{3}(\sigma_1 + \sigma_2 + \sigma_3)$$

These two equations, when rearranged, give:

$$e_1 = \frac{\sigma_1}{E}(1+v) - \frac{3v}{E}\sigma_m \qquad \text{and since} \qquad G = \frac{E}{2(1+v)}$$

$$e_1 = \frac{\sigma_1}{2G} - \frac{3v\sigma_m}{2G(1+v)} = \frac{1}{2G}\left(\sigma_1 - \frac{3v\sigma_m}{1+v}\right) \qquad (3\text{-}8)\dagger$$

Thus, the origin or *zero* of the strain circle is displaced an amount equal to $3v\sigma_m/(1+v)$ from the origin of the stress circle. Note that if v is equal to $\frac{1}{2}$, the origin of the strain circle coincides with the value of σ_m of the stress circle. (Of course, the value of v for most solids is not $\frac{1}{2}$.)

Two examples will illustrate the above findings. Consider uniaxial tension, Fig. 3-3, where $\sigma_1 = 12$ and $v = \frac{1}{3}$. Then $\sigma_m = 4$ and $3v\sigma_m/(1+v) = 3$. By using scaled plots in both circles, it is seen that e_1 is three times e_2 or e_3 in magnitude and opposite in sign. Thus, $e_2 = e_3 = -\frac{1}{3}e_1$ which must be

†Other forms are also used; see Ford [2].

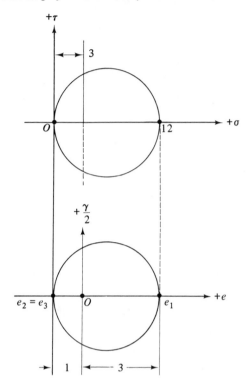

Figure 3-3 Mohr's circle of stress (a) and the corresponding strain (b) for uniaxial tension.

true since $v = \frac{1}{3}$. A physical meaning of Fig. 3-3 is that the principal stress and strain axes coincide.*

As a second example consider pure shear where $\tau_{max} = \sigma_1 = -\sigma_3$. Figure 3-4 shows three interpretations. Using Eq. (3-1):

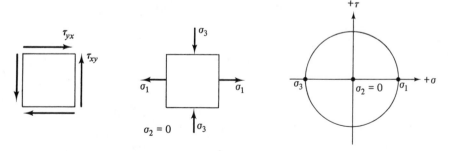

Figure 3-4 Equivalent descriptions of pure shear.

*To determine the magnitudes of the strains from Mohr's circle, an appropriate scale factor must be introduced as shown in Example 3-3.

$$e_1 = \frac{\sigma_1}{E}(1 + v), \qquad e_3 = -\frac{\sigma_1}{E}(1 + v), \qquad e_2 = 0$$

and, since $\sigma_m = 0$, $3v\sigma_m/(1 + v) = 0$.

For this situation, the centers of the stress and strain circles coincide. Note also that $\Delta = 0$ which means there is no volume change; this should be expected since pure shear produces a shape change only.

Example 3-3.

The elastic modulus of a solid is 100 GPa and it has a Poisson's ratio of $\frac{1}{3}$. It is subjected to a uniaxial tensile stress of 180 MPa. Construct the Mohr's circles for stress and corresponding strains.

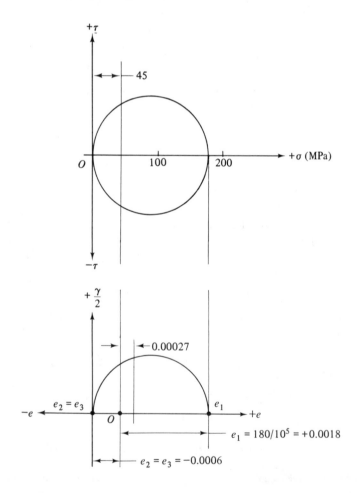

Solution.

$\sigma_m = 60$ so the origin of the strain circle is displaced from the origin of the stress circle by $3(\frac{1}{3})60/(1 + \frac{1}{3}) = 45$.

$e_2 = e_3 = -ve_1 = -0.0006$ which checks the scaled plot.

3-6 STRAIN ENERGY DUE TO THE WORK OF ELASTIC DEFORMATION

Figure 3-5(a and b) show a normal and simple shear stress acting on stress

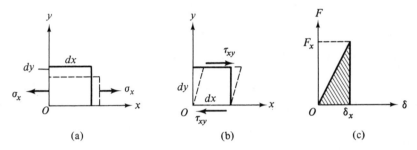

Figure 3-5 Basis for determining strain energy induced by applied stresses and accompanying strains.

elements of dimensions dx, dy, dz. In part (a), the stress σ_x produces a strain e_x and the work done is $\frac{1}{2}F_x \, \delta_x$ where:

$$F_x = \sigma_x \, dy \, dz \qquad \text{and} \qquad \delta_x = e_x \, dx \qquad \text{[see Fig. 3-5(c)]}$$

Therefore, $dw = \frac{1}{2}F_x \delta_x = \frac{1}{2}\sigma_x e_x \, dx \, dy \, dz = \frac{1}{2}\sigma_x e_x \, dv$, thus the *work per unit volume*

$$\frac{dW}{dV} = \frac{1}{2}\sigma_x e_x \tag{3-9}$$

Similarly, in Fig. 3-5(b), $dW = \frac{1}{2}(\tau_{xy} \, dx \, dz)(\gamma_{xy} \, dy) = \frac{1}{2}\tau_{xy}\gamma_{xy} \, dx \, dy \, dz$, thus:

$$\frac{dW}{dV} = \frac{1}{2}\tau_{xy}\gamma_{xy} \tag{3-10}$$

In the most general case, and with the use of superposition, the work or strain energy per unit volume is expressed by:

$$W_v = \frac{1}{2}(\sigma_x e_x + \sigma_y e_y + \sigma_z e_z + \tau_{xy}\gamma_{xy} + \tau_{yz}\gamma_{yz} + \tau_{zx}\gamma_{zx}) \tag{3-11}$$

Where principal directions are involved:

$$W_v = \frac{1}{2}(\sigma_1 e_1 + \sigma_2 e_2 + \sigma_3 e_3) \tag{3-12}$$

Different interpretations of Eq. (3-11) or Eq. (3-12) exist, as discussed in Chap. 4.

3-7 TWO IMPORTANT PHYSICAL SITUATIONS—PLANE STRESS AND PLANE STRAIN

Because they are often encountered, at least to a degree that is acceptable if not exact, the special cases of plane stress and plane strain are presented.

Plane stress (often called biaxial) occurs when

$$\sigma_z = \tau_{xz} = \tau_{yz} = 0, \quad \text{thus } z \text{ is a principal direction.}$$

Plane strain occurs when displacements are everywhere independent of one direction, that is, the strain in one direction is zero. In effect, all displacement occurs in one plane and is everywhere parallel. If, for instance, $e_z = 0$, then

$$\sigma_z = v(\sigma_x + \sigma_y) \tag{3-13}$$

This relationship is *always* true when plane strain occurs.

Finally, we again stress that an isotropic condition was assumed throughout this chapter. Anisotropy leads to more extensive relationships but these are not discussed in this text.

REFERENCES

[1] N. H. POLAKOWSKI and E. J. RIPLING, *Strength and Structure of Engineering Materials* (Englewood Cliffs, N.J.: Prentice-Hall, Inc., 1966), pp. 116–18.

[2] H. FORD, *Advanced Mechanics of Materials* (New York: John Wiley & Sons, Inc., 1963), pp. 187–89.

PROBLEMS

3-1 A thin-walled tube with closed ends is subjected to internal pressure P. The tube radius is R and wall thickness is t. The following are adequate for stress calculations:

$$\text{Hoop stress, } \sigma_1 = PR/t$$

$$\text{Axial stress, } \sigma_2 = PR/2t$$

$$\text{Radial stress, } \sigma_3 \approx 0$$

Comment on the qualitative magnitude of the strain e_2 depending on the magnitude of Poisson's ratio, where

$$0 < v < \tfrac{1}{2}, \quad v = \tfrac{1}{2}, \quad v > \tfrac{1}{2}.$$

3-2 A thin sheet of steel, $\tfrac{1}{16}$ in. thick by 10 in. wide by 24 in. long is subjected to a pure bending moment applied at the ends of the 24 in. dimension. At the instant the radius of the inner fiber is 30 in., determine the state of stress that exists at the surface. You may assume there is no shift in the neutral axis, the yield strength is 40 ksi, Poisson's ratio is 0.3 and the elastic modulus is 30 ×

10^6 psi. As a first solution, assume that the 10 in. by $\frac{1}{16}$ in. section remains rectangular throughout. Comment on this, and complete a solution where this assumption is not invoked.

3-3 Consider a cube of steel that is 1 in. per side subjected to a uniaxial tensile stress of 30 ksi.
(a) Determine the magnitude of volume increase or decrease.
(b) Repeat (a) assuming the stress is compressive rather than tensile.

3-4 A thin sheet of aluminum is subjected to the unbalanced biaxial tensile stress state, $\sigma_x = 10$ ksi, $\sigma_y = 5$ ksi, $\sigma_z = 0$ (z is the thickness direction). The thickness before loading is 0.035 in.
(a) Find the three principal strains.
(b) What is the thickness after the stresses are applied? Ignore end effects.

3-5 Under plane stress loading (say σ_x and σ_y) the strains $e_x = +0.001$ and $e_y = -0.0007$ are measured (each in inches/in.). If $E = 15 \times 10^6$ psi and $v = \frac{1}{3}$, find e_z.

3-6 A block of metal is subjected to three mutually perpendicular normal stresses of $\sigma_x = 0.14$ GPa, $\sigma_y = -0.2$ GPa, and $\sigma_z = 0.25$ GPa. If the elastic modulus is 200 GPa and Poisson's ratio is 0.3, determine the work or strain energy per unit volume of metal that is induced. Assume all deformation is elastic.

3-7 Loading on a slab of material induces plane strain deformation. If direction 1 coincides with loading, 2 with the direction of zero strain, and no loading occurs in the 3 direction, determine:
(a) the magnitudes of the three normal strains
(b) the strain energy per unit volume that is induced.
Here, $\sigma_1 = 50$ ksi, $E = 28 \times 10^6$ psi and $v = 0.33$.

3-8 The elastic modulus of a solid is 10^7 psi and $v = 0.35$. If this body is subjected to normal stresses where $\sigma_1 = 500$ MPa, $\sigma_2 = -150$ MPa and $\sigma_3 = 0$:
(a) Construct the Mohr's circle of stress.
(b) Following the procedure in Example 3-3, construct the corresponding strain circle and by introducing a proper *scale* to this circle, find the three normal strains.
(c) Check the results in part (b) by using Eq. (3-1).

3-9 A single crystal of copper is oriented so that a stress of 1000 psi (tension) acts in the [001] and [00$\bar{1}$] directions, and a stress of -1600 psi (compression) acts in the [110] and [$\bar{1}\bar{1}0$] directions. Calculate the maximum shear stress on the (111) plane. (Note that the faces of the crystal shown in the sketch are not necessarily {100} cube faces.)

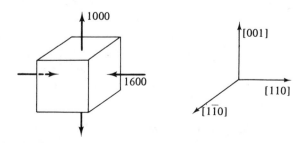

4

MACROSCOPIC PLASTICITY

4-1 INTRODUCTION

Two basic factors governed the developments in Chap. 3 on elasticity. One implied complete reversibility of the loading process; thus, when the forces that caused the body to be strained were removed, complete recovery to the initially unstrained state occurred immediately. The second factor implied that deformation or strain of the body under load was dependent only upon the end stresses and not upon the stress history or strain path. Therefore, elastic behavior may be viewed as a *point* function since any induced strain can be determined from the initial and final stresses and particular proportionality constants. The same two factors are not evident when plastic or permanent deformations are induced.

To produce plastic deformation or flow, a certain level of stress, henceforth called the *yield stress*, must be exceeded. With many solids (e.g., ductile metals) such deformation or shape change can continue to a large degree if the yield stress is greatly exceeded; in addition, as the final deformation is produced, a particular strain element could undergo different histories prior to the body reaching its end state. Thus, not only is complete reversibility not found upon load removal, as in elasticity, but the final strain is found to depend upon the history of loading rather than the beginning and final stresses alone. Such a finding means that plastic behavior is a *path* function and requires the use of *incremental* strains summed up over the strain path whenever the total induced strain is to be determined.

There are certainly at least three fairly distinct approaches that have been taken in the study of plasticity; these are:

1. The *mathematical* approach which employs idealized models of material behavior and is concerned primarily with the stress and strain distributions that satisfy prescribed boundary conditions. This might properly be called the macroscopic theory of plasticity and is most analogous to the longer standing topic known as the theory of elasticity.

2. The approach utilized in *metal physics*. Here, the manner of deformation in single crystals of real solids forms the basis of study with one objective being to relate and extend the basic behavior of single crystals to polycrystalline aggregates that make up the solids usually employed by engineers. This might properly be called the microscopic theory of plasticity.

3. The *technological* approach, which attempts to unite experimentally observed behavior of real solids on a macroscopic scale with mathematical expressions by invoking certain phenomenological rules. This enables useful predictions to be made in the general area of design; it might properly be called macroscopic engineering plasticity. It is this approach that will receive major attention in this chapter.

4-2 *COMPARISON OF ELASTICITY AND PLASTICITY*

For convenience, many of the above comments are summarized in tabular form below. In this way a direct comparison, indicating the major differences in these two types of behavior, can be seen.

Since the onset of yielding and the behavior that might follow is of primary concern, various models will be used in order to illustrate the physical processes involved. With any of the models presented below, several assumptions are invoked:

Topic	Elasticity	Plasticity
1. Stress-strain relations	Hooke's law	Flow rules
2. Equilibrium equations	Stress relations	Stress relations
3. Compatibility	Strains or displacements	Strain rates or velocities
4. Boundary conditions	Stresses and strains	Stresses and strains
5. Volume changes	Function of Poisson's ratio	None
6. Shape changes	Minimal (use total strains as point functions)	May be drastic (use incremental strains as a path function)
7. Yield condition	None	Function of stresses needed to cause plastic flow or deformation

1. The solid is isotropic and homogeneous.
2. The onset of yielding in tension and compression is identical. This implies there is no *Bauschinger effect*.*
3. Volume changes are negligible. Thus, the dilatation is zero and *Poisson's ratio* is one-half. Although this ratio is an elastic constant, no confusion should result by extending this meaning to plastic deformation.
4. The magnitude of the mean normal stress or *hydrostatic* component of the stress state does not influence yielding.
5. Effects of strain rate are negligible.
6. Temperature effects are not considered.

Note that assumptions three and four have been reasonably substantiated by experiments with many ductile metals. This is not the case with many solid polymers.

4-3 MODELS FOR PLASTIC DEFORMATION

RIGID PEREFECTLY PLASTIC SOLID

Rigid perfectly plastic behavior has found wide use in many analytical studies. It implies that no deformation occurs until a certain level of stress is reached (i.e., E is infinite), then deformation proceeds indefinitely as long as the necessary flow stress is applied. Note that nothing is implied in regard to potential fracture of the solid. A satisfactory model is shown in Fig. 4-1.

*This is discussed in Chap. 7.

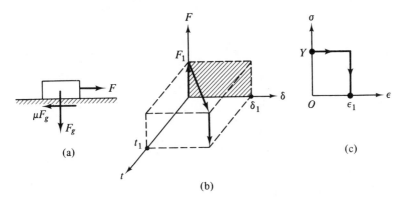

Figure 4-1 Description of a rigid perfectly plastic solid showing the model (a), the force-displacement-time plot (b), and the stress-strain plot (c).

Note the following:

1. As the applied load F is increased, no displacement occurs until some critical force F_1 is reached. Once this happens, deformation proceeds continuously with time. The force F_1 is related directly to the yield or flow stress Y.

2. Upon removal of load F_1, there is no recovery of plastic work (shown by the shaded area in the F-δ plane). Rather, a permanent deformation given by δ_1 remains.

3. The solid does not become stronger during deformation. This implies there is no *work-hardening* effect.

RIGID LINEAR WORK-HARDENING SOLID

A rigid linear work-hardening model is somewhat more realistic than the previous model since it incorporates the influence of work hardening observed in many solids, especially ductile metals. Again, a certain critical stress level must be reached before plastic deformation commences, but continued deformation demands an increasing applied stress. This is shown in Fig. 4-2. The following effects are noted:

1. As F is applied, displacement begins only when a critical force F_0 is reached and produces the initial flow stress, Y_0.

2. Displacement continues only under an increase in applied stress Y where $Y = Y_0 + f(\epsilon)$ and $f(\epsilon)$ is related to the slope of the line. Its similarity to the modulus E in Fig. 3-1(c) should be noted. In

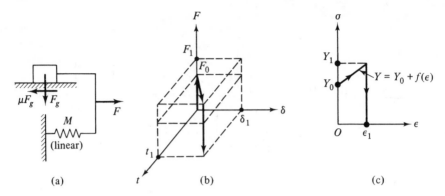

Figure 4-2 Description of a rigid linear work-hardening solid with plots
as described in Fig. 4-1.

this model, work hardening occurs and implies that plastic deformation causes an increase in the stress required for *further* deformation.

3. Upon removal of load F_1, no recovery occurs.

Rigid Nonlinear Work-Hardening Solid

A rigid work-hardening model where such hardening follows a power law
form of behavior provides an even better description for many solids. In
Chap. 5 it is shown that it is most convenient to uncouple the plastic behavior
and use the power law expression to define the effect of plastic strain on the
subsequent yield stress of the material. Figure 4-3 portrays this model. Here

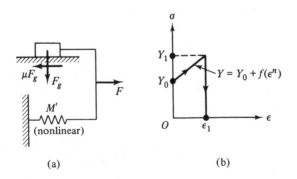

Figure 4-3 Description of a rigid work-hardening solid following a power
law hardening behavior. Plots are described in Fig. 4-1.

it is noted that:

1. The behavior is identical with the previous model except that strain hardening occurs at a nonlinear rate, the exponent n being greater than zero but less than unity.
2. The use of a power law behavior will be fully interpreted in Chap. 5.

Finally, elastic effects can be included in any of the three models by adding on a straight-line section in the initial stages of deformation where the slope would indicate an elastic modulus of an appropriate value that is less than infinity. Because many situations of concern involve plastic strains that are orders of magnitude greater than the elastic strains, it is convenient to ignore the latter. In so doing, three facts should be realized:

1. Volume changes can only be determined by including elastic effects where v is less than one-half. By ignoring such effects, the concept of volume constancy may be introduced.
2. Recovery upon loading generally does occur and is tied in with elastic recovery. Thus, if such a result is of immediate concern, the above models would not describe such behavior. Note also that in situations involving the effects of elastic recovery, continually increasing *elastic* strains accompany an ever-increasing plastic flow.
3. If elastic and plastic strains are of the same order of magnitude, the above models would not be useful *unless* the elastic portion were included as mentioned above.

4-4   THE YIELD LOCUS AND SURFACE

With the assumptions of isotropy, no Bauschinger effect, incompressibility during plastic flow and yielding being uninfluenced by hydrostatic effects, several inherent conditions must prevail in any criterion that is to be used to predict the onset of yielding.

A plot of two-dimensional *stress space* is introduced to indicate some of the results of the above assumptions. Here it is envisioned that the individual stresses may be treated as components of the total stress and are handled as vectors for this purpose. This has *nothing* to do with *transformations* to new axes and must be so understood. This discussion is also restricted to the use of principal stresses where in all cases, one of these stresses is zero. A plot in $\sigma_1 - \sigma_2$ space will be used. Suppose a tensile stress is applied in one direction and $0 < \sigma_1 < Y$ describes elastic behavior only. With equivalence in tension and compression, the elastic range is extended to $-Y < \sigma_1 < Y$ and because of isotropy, $-Y < \sigma_2 < Y$. Thus, there exist four points in $\sigma_1 - \sigma_2$ stress space that indicate the onset of yielding but to develop an acceptable theory of yielding more complex stress states must be included. This requires a generalization of what is meant by the elastic range and yield point and involves the use of certain *stress limits*. Figure 4-4 shows how this is started.

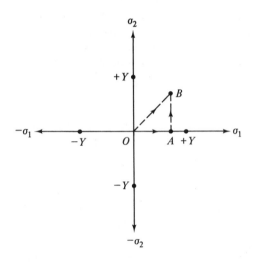

Figure 4-4 Yield points in two-dimensional stress space.

The four points shown at $\pm Y$ fall on a yield *locus* in this two-dimensional stress space. Suppose now that the material is stressed to point A as shown and that stress is maintained while a stress σ_2 is added. At some point, such as B, elastic behavior ends and we refer to B as a *yield point* in stress space. Note that the simultaneous loading in the one and two directions could have proceeded along the line OB such that yielding again occurred at B. Thus, to reach B, numerous loading paths might be followed and until that yield point is reached, all behavior is elastic. Using a number of loading paths, the locus described by the resulting yield points divides elastic behavior (inside the locus) and the onset of yielding which is the locus itself. Considering the models that included work hardening, these implied that such an effect tended to increase the subsequent yield strength or new flow stress. In this text any such tendencies will be assumed to enlarge the initial yield locus in a uniform manner; this is called isotropic hardening.

It is appropriate here to introduce the concept of three-dimensional stress space. In Fig. 4-5, the combinations of stresses a, b and c, acting in the 1, 2

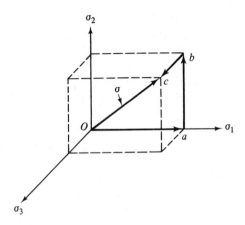

Figure 4-5 Stress resultant in three-dimensional stress space.

and 3 directions are assumed to just cause yielding. This total stress state is defined by σ which originates at the origin and its *tip* in space provides a yield point. If enough experiments were conducted, all such points would lie on a *yield surface*. Any stress state, described by a single vector such as σ, that lies within the surface causes elastic effects only. As the tip of such a vector approaches the surface, yielding is incipient. Note that a yield locus is described by passing a plane through the surface with one of the three principal stresses being a constant (e.g., $\sigma_3 = 0$ in the earlier development).

Considering that the magnitude of the mean normal stress, σ_m, does not influence yielding, the concept of a yield surface can be explained more fully. Reference to Fig. 4-6 will clarify the meaning of σ_m where an applied

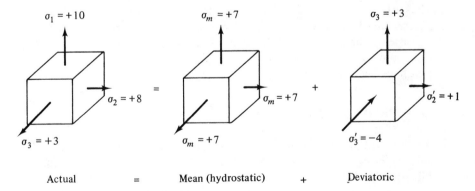

Figure 4-6 Stress state separated into mean and deviatoric components.

stress state is indicated on the left element. As shown, σ_m is equal to one third the algebraic sum of the three normal stresses; compressive stresses would count negative in this summation. If the mean stress is subtracted from each applied stress, the *deviatoric stresses* result. Literally, these deviate from the mean and if yielding is uninfluenced by σ_m then these deviatoric stresses must in some way induce yielding. It will be shown that they are nothing more than functions of shear stresses.

Thus, if a combination of stresses $(\sigma_1, \sigma_2, \sigma_3)$ just causes yielding, then $(\sigma_1 + \sigma_0, \sigma_2 + \sigma_0, \sigma_3 + \sigma_0)$ must also cause yielding and all combinations of $(\sigma_1, \sigma_2, \sigma_3)$ for various σ_0 therefore generate a line on the yield surface that is parallel to the line $\sigma_1 = \sigma_2 = \sigma_3$. This line is defined by equal direction cosines with respect to the 1, 2, 3 system. Now since isotropy and no Bauschinger effect are assumed, the rotation of such a line around the *space diagonal* $(\sigma_1 = \sigma_2 = \sigma_3)$ must generate a prism which is the yield surface. To fully specify this prism, both its cross-sectional shape and size must be defined.

All planes perpendicular to the space diagonal are defined by the equation $\sigma_1 + \sigma_2 + \sigma_3 =$ constant, this being $3\sigma_m$ for any one group of normal stresses. If the constant is set equal to zero, that plane passes through the origin at right angles to the axis of the prism; it is often called the π plane and its intersection with the yield surface is referred to as the C curve. This indicates that any point on the yield surface can be reduced to its equivalent point on the C curve simply by applying a proper value of σ_0 which simply moves the original point up or down the yield surface. Finally, consider a stress state as follows:

$$(\sigma_1, \sigma_2, \sigma_3) = (6, -2, 1) \qquad \text{and} \qquad \sigma_m = \tfrac{5}{3}$$

If this stress state is "reduced" by σ_m, then

$$(\sigma_1', \sigma_2', \sigma_3',) = (\tfrac{13}{3}, -\tfrac{11}{3}, -\tfrac{2}{3}) \qquad \text{and} \qquad \sum \sigma_i' = 0$$

Thus, the three-dimensional stress vector is composed of the deviatoric stresses that lie in the π plane and the mean or hydrostatic component that is perpendicular to the π plane. Now that the physical meaning of a yield locus and surface has been developed, we consider some possible surfaces that have been proposed.

4-5 YIELD CRITERIA

As discussed in Chap. 1, for any three-dimensional stress state there exists a cubic equation whose three roots are the principal stresses.* A useful form of this equation is:

$$\sigma_p^3 - I_1 \sigma_p^2 - I_2 \sigma_p - I_3 = 0 \tag{4-1}$$

*Ford [1] gives a detailed proof.

where the invariants, I_1, I_2, and I_3, may be expressed as functions of principal stresses as follows:

$$I_1 = (\sigma_1 + \sigma_2 + \sigma_3)$$

$$I_2 = -(\sigma_1\sigma_2 + \sigma_2\sigma_3 + \sigma_3\sigma_1) \qquad (4\text{-}2)$$

$$I_3 = (\sigma_1\sigma_2\sigma_3)$$

It may be noted immediately that $I_1 = 3\sigma_m$; thus the first invariant is a function of the hydrostatic or mean component and should not influence yielding. Therefore, any acceptable yield criterion should not include any reference to I_1 for those solids whose yield behavior has been found to be independent of σ_m.

Suppose that a yield criterion is proposed as follows: When $\sigma_1 - \sigma_2 - \sigma_3 = $ constant $= +10$, yielding will occur. If this were an acceptable criterion then $\sigma_1 = +5$, $\sigma_2 = -2$, $\sigma_3 = -3$, provides a stress state that would lead to yielding. Now superimpose a stress, $\sigma_0 = +10$. This means the new stress state is $(15, +8, +7)$ which, according to the proposed criterion, would not cause yielding since $\sigma_1 - \sigma_2 - \sigma_3$ is zero and not $+10$. Yet only the mean stress was varied. Such a criterion would not agree with experimental observations so it could not be considered to possess the generality required. The two most widely used criteria both satisfy independence of I_1† and have found best agreement when experiments have utilized ductile metals. Each criterion is associated with numerous names in the literature; the interested reader may check the work of Paul [2] for a thorough historical perspective.

4-6 TRESCA CRITERION

This criterion [3] proposes that yielding will occur when some function of the *maximum* shear stress reaches a critical value. Whenever possible, the convention $\sigma_1 > \sigma_2 > \sigma_3$ will be used but there are cases when this relative comparison is not known a priori. In addition, this convention cannot be maintained rigorously when plots in two- or three-dimensional stress space are considered.

It is useful to recall that when the three Mohr's circles are plotted it is the radius of the largest circle that gives the maximum shear stress. Accounting for algebraic signs, this criterion is written as:

$$|\sigma_{max} - \sigma_{min}| = \text{constant} = |\sigma_1 - \sigma_3| \qquad \text{if } \sigma_1 > \sigma_2 > \sigma_3 \qquad (4\text{-}3)$$

†Reduced stress invariants are also used. They result by subtracting $I_1/3$ from each stress. In terms of principal stresses, the reduced invariants are: $J_1 = 0$, $J_2 = \frac{1}{6}[\sigma_1 - \sigma_2)^2 + (\sigma_2 - \sigma_3)^2 + (\sigma_3 - \sigma_1)^2]$, and $J_3 = \frac{1}{27}(2I_1^3 + 9I_1I_2 + 27I_3)$, which is too cumbersome to expand in terms of stresses.

If such a criterion finds reasonably universal acceptance *regardless* of the applied stress state, then the constant should be readily determined from simple standard tests.

(a)   For uniaxial tension, yielding occurs when σ_1 reaches the uniaxial yield stress, Y. Thus:

$$\sigma_1 = Y, \qquad \sigma_2 = \sigma_3 = 0 \qquad \text{and} \qquad \tau_{max} = \tfrac{1}{2}Y$$

Using Eq. (4-3), this means:

$$|\sigma_1 - 0| = \text{constant} = Y$$

(b)   For pure shear, $\sigma_1 = -\sigma_3 = \tau_{max}$, $\sigma_2 = 0$. For convenience, let the maximum allowable shear stress be designated as k, the *shear* yield stress. Using Eq. (4-3), we get:

$$|\sigma_1 - (-\sigma_1)| = \text{constant} = 2\sigma_1 = 2k.$$

Thus, the Tresca criterion may be expressed as:

$$|\sigma_{max} - \sigma_{min}| = Y = 2k = |\sigma_1 - \sigma_3| \qquad \text{if } \sigma_1 > \sigma_2 > \sigma_3 \qquad (4\text{-}4)$$

If a solid obeyed this criterion exactly, then the tensile and shear yield stresses would relate in a two-to-one ratio. This does not mean this ratio must be observed; rather, it is *predicted* by this criterion.

Example 4-1.

A material whose tensile yield strength, Y, is 50 ksi is subjected to a uniaxial compressive stress of 30 ksi. Determine the magnitude of the tensile stress applied at right angles to the initial compressive stress, that would cause yielding according to the Tresca criterion. Plot the Mohr's circle for this situation.

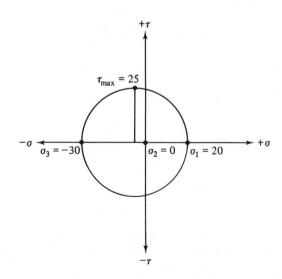

Solution.

$$|\sigma_1 - \sigma_3| = Y \qquad \text{where } \sigma_1 > \sigma_2 > \sigma_3$$

Here σ_2 is indicated as zero, σ_3 is negative, and σ_1, the unknown stress, is positive.

$$|\sigma_1 - (-30)| = 50, \qquad \text{so } \sigma_1 = 20 \text{ ksi}$$

Note that the diameter of the circle is 50 ksi.

4-7 VON MISES CRITERION

Perhaps because the Tresca criterion produces a yield surface having corners on the prism that results in stress space and because the intermediate stress, σ_2, is ignored, a criterion that involves a smooth function was proposed by von Mises [4]. Although the mathematical statement reduces to $6J_2 = $ constant, this gives little concrete understanding. Several physical interpretations have been suggested; these are called the distortion energy and octahedral shear stress theories. It is to be emphasized that they are interpretations which followed the proposed mathematical postulate, but there is no point in indulging in historical sequencing as far as the use of this criterion is concerned. In its most widely used form, the von Mises criterion, in terms of principal stresses, predicts that yielding occurs when

$$(\sigma_1 - \sigma_2)^2 + (\sigma_2 - \sigma_3)^2 + (\sigma_3 - \sigma_1)^2 = \text{constant} \qquad (4\text{-}5)^*$$

In a more general form:

$$\begin{aligned}(\sigma_x - \sigma_y)^2 + (\sigma_y - \sigma_z)^2 + (\sigma_z - \sigma_x)^2 \\ + 6(\tau_{xy}^2 + \tau_{yz}^2 + \tau_{zx}^2) = \text{constant}\end{aligned} \qquad (4\text{-}6)$$

The proof of the equivalence of these equations constitutes an exercise for the interested reader. With Eq. (4-5) there is no need to know at the outset how the principal stresses relate algebraically since all are equally weighted. Note also that each stress difference is a *shear stress* connected with one of the three Mohr's circles, so as with the Tresca criterion, shear stresses tie in with the onset of yielding. To determine the constant, the same procedure used in Sec. 4-6 is used.

(a) For uniaxial tension, yielding occurs when $\sigma_1 = Y, \sigma_2 = \sigma_3 = 0$.
Using Eq. (4-5):

$$2\sigma_1^2 = \text{constant} = 2Y^2$$

(b) For pure shear, yielding occurs when $\sigma_1 = -\sigma_3 = k, \sigma_2 = 0$ and
using Eq. (4-5):

$$\sigma_1^2 + \sigma_1^2 + 4\sigma_1^2 = \text{constant} = 6\sigma_1^2 = 6k^2$$

*Note the similarity with $6J_2$ in the footnote on page 67.

Thus, the von Mises criterion may be written as:

$$(\sigma_1 - \sigma_2)^2 + (\sigma_2 - \sigma_3)^2 + (\sigma_3 - \sigma_1)^2 = 2Y^2 = 6k^2 \quad (4\text{-}7)$$

According to this criterion, the tensile and shear yield stresses are related as,

$$Y = \sqrt{3}\,k$$

which is the first inkling that the two criteria may lead to different predictions.

It is convenient to consider each criterion as a function of an *effective stress* denoted as $\bar{\sigma}$ where $\bar{\sigma}$ is a function of the applied stresses. Whenever its *magnitude* reaches the yield strength in *uniaxial tension*, then that applied stress state should cause yielding to occur (i.e., it has reached an effective level). Thus:

von Mises: $\quad \bar{\sigma} = \dfrac{1}{\sqrt{2}}[(\sigma_1 - \sigma_2)^2 + (\sigma_2 - \sigma_3)^2 + (\sigma_3 - \sigma_1)^2]^{1/2} \quad (4\text{-}8)$

Tresca: $\quad\quad\quad\quad\quad \bar{\sigma} = |\sigma_{max} - \sigma_{min}| \quad\quad\quad\quad\quad\quad\quad (4\text{-}9)$

When $\bar{\sigma}$ reaches a value of Y because of the effects of σ_1, σ_2, and σ_3 either criterion predicts yielding. However, according to the von Mises criterion when $\bar{\sigma}$ reaches a value of $\sqrt{3}\,k$ yielding is predicted whereas the Tresca criterion requires $\bar{\sigma}$ to reach $2k$ before yielding is expected.

There are many ways these two rules can be expressed but it is of importance to realize they are not natural laws as such and their initial postulation was due to mathematical rather than physical reasoning. It is a bit surprising that these two postulates have provided such close agreement with physical observations.

Example 4-2.

Repeat Example 4-1 using the von Mises criterion for predictive purposes. Using Eq. (4-7):

Solution.

$$[(\sigma_1 - 0)^2 + (0 - (-30)]^2 + (-30 - \sigma_1)^2 = 2(50)^2$$

$$\sigma_1^2 + 900 + 900 + 60\sigma_1 + \sigma_1^2 = 5000$$

$$\sigma_1^2 + 30\sigma_1 - 1600 = 0$$

$$\sigma_1 = \frac{-30 \pm (900 + 6400)^{1/2}}{2} = \frac{-30 \pm 85.44}{2}$$

so,

$$\sigma_1 = -57.72 \text{ or } +27.72 \text{ ksi.}$$

In view of the question the tensile value of 27.72 ksi is the correct answer. The negative value of -57.72 ksi is the compressive stress that is required to cause yielding if such a loading were of concern. Note that τ_{max} equals $Y/\sqrt{3}$ here whereas τ_{max} equals $Y/2$ in the previous example.

That neither criterion is influenced by σ_m can be shown by reference to Fig. 4-7 where the deviatoric stresses (as functions of shear stresses) are seen to be crucial. Suppose that upon the original stress state, a stress component of $+9$ is superimposed, then $\sigma_1 = 48$, $\sigma_2 = 17$, $\sigma_3 = 13$ and the new $\sigma_m = 26$. The center of the largest circle is at $\sigma = 30.5$, and $\tau_{max} = 17.5$. Note that $\sigma_1' = 22$, $\sigma_2' = -9$, and $\sigma_3' = -13$, as above! Thus, the change in σ_m displaces the circles but neither changes their size nor influences the magnitudes of the deviatoric stresses. Note too that:

$$\sigma_1' = \sigma_1 - \sigma_m = \frac{(\sigma_1 - \sigma_2) + (\sigma_1 - \sigma_3)}{3}$$

(and similarly for σ_2' and σ_3'), so the deviatoric stresses are functions of shear stresses.

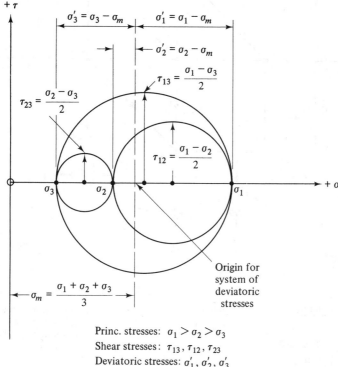

Princ. stresses: $\sigma_1 > \sigma_2 > \sigma_3$
Shear stresses: $\tau_{13}, \tau_{12}, \tau_{23}$
Deviatoric stresses: $\sigma_1', \sigma_2', \sigma_3'$

Suppose $\sigma_1 = 39$, $\sigma_2 = 8$, and $\sigma_3 = 4$ (all in ksi), then $\sigma_m = 17$, center of largest circle at $\sigma = 21.5$, and $\tau_{max} = 17.5$. Note that $\sigma_1' = 22$, $\sigma_2' = -9$, and $\sigma_3' = -13$.

Figure 4-7 Mohr's circle for three-dimensional stress state showing principal stresses, mean normal stress, and deviatoric components.

Many practical problems can be approximated by or reduced to a biaxial (plane) stress situation. Consider that $\sigma_2 = 0$. If a plot is made in $\sigma_1 - \sigma_3$ stress space, Fig. 4-8 results for the Tresca criterion and Fig. 4-9 shows the Mises plot. If the two plots are combined, Fig. 4-10 results. The various con-

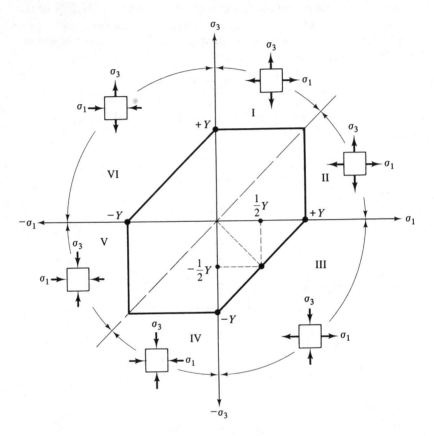

Zone	Condition	Boundary
I	$\sigma_3 > \sigma_1 > 0$	$\sigma_3 = +Y$
II	$\sigma_1 > \sigma_3 > 0$	$\sigma_1 = +Y$
III	$\sigma_1 = -\sigma_3 = k = Y/2$	45° line as shown
IV	$0 > \sigma_1 > \sigma_3$	$\sigma_3 = -Y$
V	$0 > \sigma_3 > \sigma_1$	$\sigma_1 = -Y$
VI	See III above	

Any stress state for σ_1, σ_3 that lies within the heavy lines (boundary) above, will not cause yielding.

Figure 4-8 Tresca yield locus.

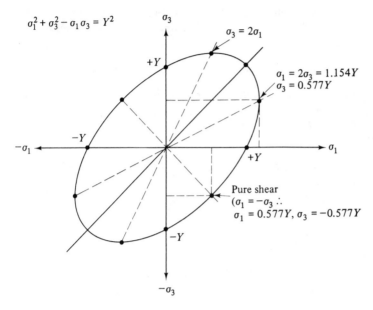

$$\sigma_1^2 + \sigma_3^2 - \sigma_1\sigma_3 = Y^2$$

$\sigma_3 = 2\sigma_1$

$\sigma_1 = 2\sigma_3 = 1.154Y$
$\sigma_3 = 0.577Y$

Pure shear
$(\sigma_1 = -\sigma_3 \therefore$
$\sigma_1 = 0.577Y, \ \sigma_3 = -0.577Y$

Figure 4-9 von Mises yield locus.

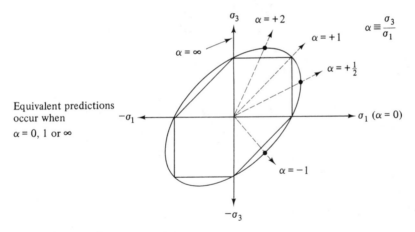

$\alpha = +2$

$\alpha = \infty$

$\alpha = +1$

$\alpha \equiv \dfrac{\sigma_3}{\sigma_1}$

$\alpha = +\frac{1}{2}$

Equivalent predictions
occur when
$\alpha = 0, 1$ or ∞

$\sigma_1 \ (\alpha = 0)$

$\alpha = -1$

Figure 4-10 Yield loci for the Tresca and von Mises criteria showing certain loading paths.

stant stress ratio loading paths in Fig. 4-10 show that the maximum predicted difference between these criteria occurs when $\alpha = -1$ (pure shear) or $\alpha = \frac{1}{2}$ or 2 (a 2:1 stress ratio). It is often overlooked that the two criteria coincide at particular points, wherein the ellipse circumscribes the hexagon, *only* because each was defined to predict yielding when the uniaxial tensile or compressive yield stress was used to define the constants in Eqs. (4-8) and (4-9).

Example 4-3.

Using a Tresca yield locus, plotted in $\sigma_1 - \sigma_2$ space, indicate the following loading sequence:
(a) $\sigma_1 = +\sigma_2$ up to a value of $Y/2$.
(b) $\sigma_1 = -\frac{2}{3}Y$
(c) $\sigma_2 = -Y$
(d) $\sigma_1 = 2\sigma_2$ where both are negative and the loading is continued until yielding is predicted.

Solution.

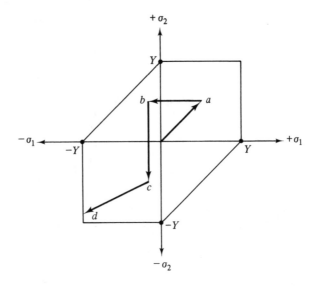

4-8 DISTORTION ENERGY

One interpretation of the Mises criterion is that yielding occurs when the elastic energy causing distortion reaches a critical value. This strain energy is found in a general way by subtracting the dilatational strain energy from the total elastic strain energy. The developments leading to Eqs. (3-11) and (3-12) showed the total strain energy per unit volume to be:

$$W_v = \tfrac{1}{2}(\sigma_x e_x + \sigma_y e_y + \sigma_z e_z + \tau_{xy}\gamma_{xy} + \tau_{yz}\gamma_{yz} + \tau_{zx}\gamma_{zx}) \qquad (4\text{-}10)$$

or for the case of principal stresses:

$$W_v = \tfrac{1}{2}(\sigma_1 e_1 + \sigma_2 e_2 + \sigma_3 e_3) \tag{4-11}$$

To express Eq. (4-11) as a function of stresses, the generalized form of Hooke's law given by Eq. (3-1) is used to give:

$$W_v = \frac{1}{2E}(\sigma_1^2 + \sigma_2^2 + \sigma_3^2) - \frac{v}{E}(\sigma_1\sigma_2 + \sigma_2\sigma_3 + \sigma_3\sigma_1) \tag{4-12}$$

Since only normal stresses cause a volume change, the dilatation is:

$$\Delta = e_1 + e_2 + e_3 = \frac{1-2v}{E}(\sigma_1 + \sigma_2 + \sigma_3) = \frac{3}{E}(1-2v)\sigma_m \tag{4-13}$$

Now the normal strains associated with σ_m must be equivalent and since $\Delta = 3e_m$:

$$e_m = \frac{(1-2v)}{E}\sigma_m \tag{4-14}$$

Observing that the work due to dilatation, $W_d = 3(\tfrac{1}{2}\sigma_m e_m)$, then $W_d = [3(1-2v)\sigma_m^2]/2E$ and finally,

$$W_d = \frac{1-2v}{6E}(\sigma_1 + \sigma_2 + \sigma_3)^2 \tag{4-15}$$

By subtracting Eq. (4-15) from (4-12) to give the shear strain energy, W_s, the result after much manipulation is:

$$W_s = \frac{1}{12G}[(\sigma_1 - \sigma_2)^2 + (\sigma_2 - \sigma_3)^2 + (\sigma_3 - \sigma_1)^2] \tag{4-16}$$

Now the shear strain energy induced during uniaxial tension where $\sigma_2 = \sigma_3 = 0$ is:

$$W_s = \frac{\sigma_1^2}{6G} \tag{4-17}$$

The critical value, W_{sc}, that must be developed to cause yielding will result when $\sigma_1 = Y$. Setting Eq. (4-16) equal to this critical value leads to:

$$\frac{1}{12G}[(\sigma_1 - \sigma_2)^2 + (\sigma_2 - \sigma_3)^2 + (\sigma_3 - \sigma_1)^2] = \frac{Y^2}{6G} \tag{4-18}$$

which is identical to Eq. (4-7). This explains why the Mises criterion is often called the distortion energy theory whose physical meaning is that yielding will occur when the elastic energy causing distortion reaches a critical value.

4-9 OCTAHEDRAL SHEAR STRESS

A second physical interpretation of the von Mises criterion has also been proposed. For simplicity consider a coordinate system defined by principal directions and a line from the origin having direction cosines where $k =$

$l = m$. The planes normal to this line and equivalent lines in other regions of space are called octahedral planes where the intersections of these eight equivalent planes form an octahedron. For this physical situation, it was shown earlier by Eq. (1-5) that:

$$S_n = \sigma_1 k^2 + \sigma_2 l^2 + \sigma_3 m^2 \tag{4-19}$$

Since $k = l = m = \cos 54° 44' = 1/\sqrt{3}$, $S_n = \frac{1}{3}(\sigma_1 + \sigma_2 + \sigma_3)$. Thus the stress *normal* to the octahedral planes is σ_m, and since σ_m has no influence upon yielding, it has been proposed that the shear stresses acting on this plane (τ_0) must reach a critical value for yielding to occur. Although not fully derived here, this stress can be shown to be:

$$\tau_0 = \frac{1}{3}[(\sigma_1 - \sigma_2)^2 + (\sigma_2 - \sigma_3)^2 + (\sigma_3 - \sigma_1)^2]^{1/2} \tag{4-20}$$

and a comparison with Eq. (4-8) shows that:

$$\tau_0 = \frac{\sqrt{2}}{3}\bar{\sigma} \tag{4-21}$$

With the use of a proper multiplying factor, this is just another version of the Mises criterion. Throughout the remainder of this text, Eq. (4-8) will be used whenever this particular criterion is to be employed.

4-10 FLOW RULES OR PLASTIC STRESS-STRAIN RELATIONSHIPS

Just as the generalized Hooke's law in the elastic region can be expressed by:

$$e_1 = \frac{1}{E}[\sigma_1 - \nu(\sigma_2 + \sigma_3)]$$

analogous relations for the plastic region are needed.

These *flow rules* can be developed in a simple way as follows, recalling that path or history dependency requires the use of incremental strains during plastic deformation.

Consider plastic flow under uniaxial tension as indicated in Fig. 4-11.

$$\sigma_m = \frac{\sigma_1 + \sigma_2 + \sigma_3}{3} = \frac{1}{3}\sigma_1$$

Now the deviatoric stress in the 1 direction is $\sigma_1' = \sigma_1 - \sigma_m$ and at the particular instant represented in Fig. 4-11,

$$\sigma_1' = \tfrac{2}{3}\sigma_1 \quad \text{and} \quad \sigma_2' = \sigma_3' = 0 - \tfrac{1}{3}\sigma_1 = -\tfrac{1}{3}\sigma_1$$

so,

$$\sigma_1' = -2\sigma_2' = -2\sigma_3'$$

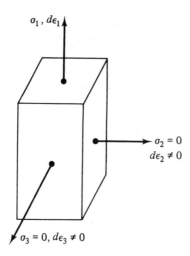

Figure 4-11 Stresses and incremental plastic strains for uniaxial tension.

For volume constancy, the sum of the plastic strain increments must be zero; therefore:

$$d\epsilon_1 + d\epsilon_2 + d\epsilon_3 = 0*$$

and because of symmetry in this instance $d\epsilon_2 = d\epsilon_3$. Therefore, $d\epsilon_1 = -2d\epsilon_2 = -2d\epsilon_3$. This leads to: $d\epsilon_1/d\epsilon_2 = -2 = \sigma'_1/\sigma'_2$, and so forth, which can be written as:

$$\frac{d\epsilon_1}{\sigma'_1} = \frac{d\epsilon_2}{\sigma'_2} = \frac{d\epsilon_3}{\sigma'_3} = \text{constant} = d\lambda \qquad \text{(in general)} \qquad (4\text{-}22)$$

(i.e., the constant isn't always -2 but these *ratios* are always in some constant proportion).

The implication is that the ratio of the *current incremental plastic strain increments* to the *current deviatoric stresses* is a constant; nothing is implied regarding the magnitudes of either. Several comments are worth noting:

1. The above development uses a simple method that produces the flow rules. In effect, it may be viewed as a necessary but not sufficient condition since no real proof is offered to indicate that other stress states would lead to the same finding.
2. Equation (4-22) expresses the Prandtl-Reuss flow rules where the elastic strain increments have been omitted.

*Solve Prob. 2-2 to be convinced.

3. Equation (4-22) is identical to the Levy-Mises equations (proposed earlier than Prandtl) where the *total* strain increment is assumed to be equivalent to the plastic strain increment. Thus, the Levy-Mises equations may be viewed as a special form of the more general expression. For a more complete explanation, see Johnson and Mellor [5].

For greater convenience, the flow rules may be expressed in various forms other than Eq. (4-22). These are:

(a) $\dfrac{d\epsilon_1 - d\epsilon_2}{\sigma_1 - \sigma_2} = d\lambda$ (4-23)

(b) $d\epsilon_1 = \frac{2}{3} d\lambda[\sigma_1 - \frac{1}{2}(\sigma_2 + \sigma_3)]$ (4-24)

(c) $d\epsilon_1 = \dfrac{d\bar{\epsilon}}{\bar{\sigma}}[\sigma_1 - \frac{1}{2}(\sigma_2 + \sigma_3)]$ (4-25)

Note the great similarity between Eq. (4-25) and the generalized Hooke's law, Eq. (3-1), where $1/E$ is replaced by $d\bar{\epsilon}/\bar{\sigma}$ and v is replaced by $\frac{1}{2}$ as a consequence of incompressibility. The coefficient $d\bar{\epsilon}/\bar{\sigma}$ is not a constant in the sense that E is; rather it is a variable proportionality factor. The incremental effective strain, $d\bar{\epsilon}$, requires definition and the form to be used from this point on is:

$$d\bar{\epsilon} = \frac{\sqrt{2}}{3}[(d\epsilon_1 - d\epsilon_2)^2 + (d\epsilon_2 - d\epsilon_3)^2 + (d\epsilon_3 - d\epsilon_1)^2]^{1/2} \quad \text{(4-26)\dag}$$

Note its great similarity with the effective stress function given by Eq. (4-8). The coefficient $\sqrt{2}/3$ is chosen so that the value of $d\bar{\epsilon}$ is equal to $d\epsilon_1$ under uniaxial tension in the same way that $\bar{\sigma}$ was made equivalent to σ_1 by the use of coefficient $1/\sqrt{2}$ in Eq. (4-8). It is crucial to realize that from Eqs. (4-8) and (4-26), *both* $\bar{\sigma}$ and $d\bar{\epsilon}$ are *always* positive, so their ratio in Eq. (4-25) must also be positive.

Flow rules for any yield criterion may be derived by using the concept of a plastic potential [6]. This method proposes that the incremental strains resulting from a stress σ_{ij} are found by using:

$$d\epsilon_{ij} = \frac{\partial f}{\partial \sigma_{ij}}(d\lambda') \quad \text{(4-27)}$$

where f is taken as the yield function. If the Mises criterion is used,

$$f(\sigma_{ij}) = (\sigma_1 - \sigma_2)^2 + (\sigma_2 - \sigma_3)^2 + (\sigma_3 - \sigma_1)^2 = \text{constant}$$

then

\dag If the three plastic strains, ϵ_1, ϵ_2, and ϵ_3 are known, the integral of $d\bar{\epsilon}$ gives the effective plastic strain $\bar{\epsilon}$. In terms of total strains, Eq. (4-26) can also be expressed as $\bar{\epsilon} = [\frac{2}{3}(\epsilon_1^2 + \epsilon_2^2 + \epsilon_3^2)]^{1/2}$.

$$\frac{\partial f}{\partial \sigma_1} = 2(\sigma_1 - \sigma_2) - 2(\sigma_3 - \sigma_1) = 4\sigma_1 - 2(\sigma_2 + \sigma_3)$$

but

$$3\sigma_m = \sigma_1 + \sigma_2 + \sigma_3$$

so

$$(\sigma_2 + \sigma_3) = 3\sigma_m - \sigma_1$$

Now,

$$\frac{\partial f}{\partial \sigma_1} = 6(\sigma_1 - \sigma_m) = 6\sigma_1'$$

Finally,

$$d\epsilon_1 = 6\sigma_1' \, d\lambda' \qquad \text{or} \qquad \frac{d\epsilon_1}{\sigma_1'} = d\lambda \qquad (4\text{-}28)$$

as in Eq. (4-22). Flow rules associated with the Tresca criterion have found little use.

It was earlier stated that Eq. (4-22) expresses a reduced form of the Prandtl-Reuss equations where the elastic strains were omitted. To clarify this point, the following illustration is given. Loading is by uniaxial tension and the normal strain in the direction of loading is to be determined.

$$(d\epsilon)_{\text{total}} = (d\epsilon)_{\text{plastic}} + (d e)_{\text{elastic}}$$

$$d\epsilon_{t1} = d\epsilon_1 + de_1$$

$$d\epsilon_{t1} = \frac{d\bar{\epsilon}}{\bar{\sigma}}\left[\sigma_1 - \frac{1}{2}(\sigma_2 + \sigma_3)\right] + \frac{1}{E}[d\sigma_1 - v(d\sigma_2 + d\sigma_3)] \qquad (4\text{-}29)$$

Note that plastic strains are related to total stresses whereas elastic strains are associated with incremental stress changes. Figure 4-12 demonstrates this point. In Fig. 4-12(a) an equal incremental change in stress causes the

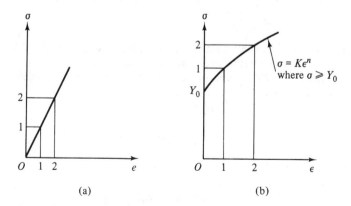

(a)   (b)

Figure 4-12 Stress-strain curves for elastic and plastic deformation.

same incremental change in strain, since a linear relation exists. However, in Fig. 4-12(b) an equal incremental stress change does not produce the same incremental change in strain; instead, these are related to a changing slope, so a particular increment depends upon the total stress at a given instant.

Example 4-4.

Consider a thin-walled tube having closed ends that is internally pressurized to 1000 psi; it is assumed that this causes plastic deformation to occur and end effects are ignored. The total strain in the hoop direction is to be found. With this loading state, the following stress relationships are acceptable.

$$\sigma_\theta = \sigma_1 = \frac{Pr}{t}, \qquad \sigma_l = \sigma_2 = \frac{Pr}{2t}, \qquad \sigma_r = \sigma_3 = 0. \qquad (a)$$

Consider the r/t ratio to be 20 and note that $\sigma_2 = \sigma_1/2 = \sigma_m$!

Solution.

With the use of Eq. (4-25),

$$d\epsilon_1 = -d\epsilon_3, \qquad d\epsilon_2 = 0 \qquad (b)$$

In incremental form, the total strain in the "1" direction is:

$$d\epsilon_t = d\epsilon + de \qquad \text{(i.e., plastic plus elastic)}$$

Using Eqs. (4-25) and (3-1):

$$d\epsilon_t = \frac{d\bar{\epsilon}}{\bar{\sigma}}\left[\sigma_1 - \frac{1}{2}(\sigma_2 + \sigma_3)\right] + \frac{1}{E}[d\sigma_1 - v(d\sigma_2 + d\sigma_3)]$$

which reduces to:

$$d\epsilon_t = \frac{d\bar{\epsilon}}{\bar{\sigma}}\left[\frac{3}{4}\sigma_1\right] + \frac{d\sigma_1}{E}\left[1 - \frac{v}{2}\right]$$

using the relations in Eq. (a). Substituting the stress relations of Eq. (a) into Eq. (4-8) and the strain relations of Eq. (b) into Eq. (4-26) gives:

$$\bar{\sigma} = \frac{\sqrt{3}}{2}\sigma_1 \qquad \text{and} \qquad d\bar{\epsilon} = \frac{2}{\sqrt{3}}d\epsilon_1 \qquad (c)$$

For the values in the problem statement, $\sigma_1 = 20$ ksi so $\bar{\sigma} = \sqrt{3}(10)$ ksi. What is now essential is an effective stress–effective strain relationship! *Assume* for now that the form $\bar{\sigma} = K\bar{\epsilon}^n$ is appropriate for the plastic portion of deformation, where $K = 25$ ksi and $n = 0.25$. Then the relation:

$$10\sqrt{3} = 25(\bar{\epsilon})^{0.25}$$

results and

$$\bar{\epsilon} = (0.693)^4 = 0.23$$

which is, from Eq. (4-26), $\int_0^\epsilon d\bar{\epsilon}$. Therefore,

$$\int_0^{\epsilon_1} d\epsilon_1 = \int_0^\epsilon \frac{\sqrt{3}}{2} d\bar{\epsilon}$$

so

$$\epsilon_1 = 0.199.$$

To compute the elastic portion of strain, note that $d\sigma_1$ equals $20\,dP$ where dP equals 1000 psi. For aluminum (whose values of K and n are reasonably well represented by the numbers used above) take $E = 10^7$ psi and $v = \frac{1}{3}$. Since $de_1 = d\sigma_1(1 - v/2)/E$, $e_1 \approx 0.002$ so $\epsilon_t = 0.199 + 0.002 = 0.201$.

Several points are pertinent.

1. Where large *plastic* strains are encountered, ignoring elastic strains introduces little error and often greatly simplifies an analysis.
2. Many real problems must invoke the use of approximations since the resulting deformation does not follow a simple loading path as used in this example.
3. Some type of $\bar{\sigma} - \bar{\epsilon}$ relationship must be available if numerical answers are required.
4. The lower limits on the above integrals were taken as zero. Physically this implies that the pressure was zero at the outset and no elastic or plastic strains had been induced in the material *prior* to the application of pressure.

4-11 NORMALITY AND YIELD SURFACES

A physical interpretation of the flow rules is that the axes of principal stresses and strains coincide. To clarify this point, it is useful to consider these components as vectors which may be plotted in three-dimensional space (as a yield surface plot) or in two-dimensional space (as a yield locus plot). It is again emphasized that this approach has nothing to do with tensor transformations; rather, the direction and magnitude of each component of stress or strain define a vector in regard to the plots that are developed.

Several approaches to the subject of normality have been presented which lead to the conclusion that the total strain vector must be normal to the yield surface. The most rigorous proof is attributed to Drucker [7] and is based upon the concept that plastic work is positive. As a consequence, any acceptable yield surface must be convex around its origin, or a straight line passing through the surface cannot cut it at more than two points.

Two simple physical examples* will assist here. First consider Fig. 4-13. Assume σ_1 is a principal stress and the material is isotropic. Two possible shape changes are shown above. On the basis of intuition alone, the change shown in (a) is far more likely to occur than that described in (b). Therefore, the stress and strain directions would coincide.

*Neither example constitutes a proof.

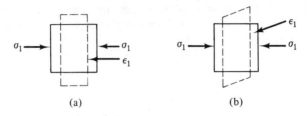

Figure 4-13 Possible conditions illustrating directions of principal stress and strain.

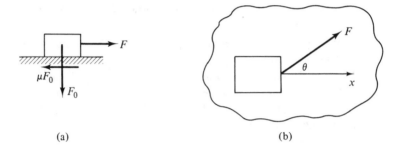

Figure 4-14 An illustration of the principle of maximum work.

Now consider a block resting on a surface as shown in Fig. 4-14 where a coefficient of friction prevails at the interface as indicated in part (a). The top view (b) indicates F acting at some angle θ with respect to the block. As F must overcome the friction force, the maximum work will result when θ reduces to zero in regard to motion of the block. Thus, it would not be expected that the block would move in the x direction and the resulting work would be maximized when the force and motion have the same direction. This is a simple example of the principle of maximum work which results in a maximum dissipation of energy, thereby lowering the total energy of the system. (In a thermodynamic sense, the external system of stresses does work which is considered positive in the usual sense.) A corollary might state that the material being deformed offers maximum resistance to plastic deformation. For this to occur, the strain vector must be normal to the yield surface.

It is now prudent to indicate, in reasonable detail, a physical development of a yield surface in three dimensional stress space. The coordinate system relates to principal directions and is shown in Fig. 4-15. The state of stress at P is caused by the principal stresses where $\sigma_1 = OP_1, \sigma_2 = OP_2, \sigma_3 = OP_3$ and OP is the total state of stress in this stress space.

In this coordinate system, consider a line OH having the same direction cosines (i.e., $k = l = m = 1/\sqrt{3}$) such that $\alpha = \beta = \gamma = 54° 44'$.

Project OP on OH to N such that angle $ONP = 90°$; then

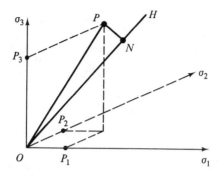

Figure 4-15 Three-dimensional stress space showing total stress OP. ON is related to hydrostatic effects whereas NP relates to the deviatoric stresses.

$$ON = \frac{1}{\sqrt{3}}(\sigma_1 + \sigma_2 + \sigma_3)$$

$$(NP)^2 = (OP)^2 - (ON)^2 = (\sigma_1^2 + \sigma_2^2 + \sigma_3^2) - [(\sigma_1 + \sigma_2 + \sigma_3)/\sqrt{3}]^2$$

$$= \sigma_1^2 + \sigma_2^2 + \sigma_3^2 - \tfrac{1}{3}[\sigma_1^2 + \sigma_2^2 + \sigma_3^2 + 2(\sigma_1\sigma_2 + \sigma_2\sigma_3 + \sigma_3\sigma_1)]$$

$$= \tfrac{2}{3}(\sigma_1^2 + \sigma_2^2 + \sigma_3^2) - \tfrac{2}{3}(\sigma_1\sigma_2 + \sigma_2\sigma_3 + \sigma_3\sigma_1)$$

$$= \tfrac{1}{3}[(\sigma_1 - \sigma_2)^2 + (\sigma_2 - \sigma_3)^2(\sigma_3 - \sigma_1)^2]$$

With the von Mises criterion, yielding occurs when $\sum (\sigma_1 - \sigma_2)^2 = 2Y^2$, so

$$3(NP)^2 = 2Y^2 \qquad \text{or} \qquad NP = Y\sqrt{2/3}$$

Thus, the yield locus can be expressed as a circle of radius $NP = Y\sqrt{2/3}$ and in principal stress space this is a circular cylinder of radius $Y\sqrt{2/3}$ whose axis is a line passing through the origin and equally inclined to the three coordinate axes.

Figure 4-16 shows the three-dimensional plot of both the von Mises "cylinder" and the Tresca "hexagon." Note that the plane where $\sigma_2 = 0$ cuts the surfaces results in the more familiar yield loci of these two criteria. By superimposing the principal strains upon the same coordinate system, the *vector sum* of the incremental strains is shown as $d\epsilon_v$; it is this quantity that is *normal* to the yield surface. It is always composed of components $d\epsilon_1$, $d\epsilon_2$ and $d\epsilon_3$ which have the same axes as σ_1, σ_2, and $\sigma\epsilon$.† Thus, the *direction* of the total strain vector, $d\epsilon_v$, depends only upon the *shape* of the yield surface. So long as P falls within the surface, yielding will not occur, but as P approaches the yield surface, flow becomes imminent. Note that ON represents the *hydrostatic component* and moving along OH (i.e., increasing or decreasing σ_m) has no influence on yielding. NP represents the *deviatoric stress* which governs yielding.

†In particular situations one or more of these six components may be zero.

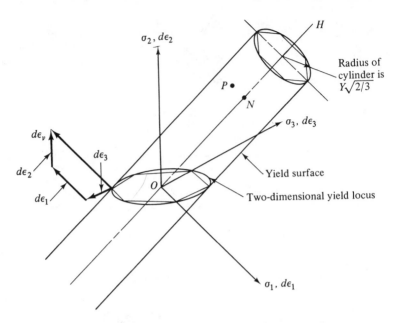

Figure 4-16 Tresca and von Mises yield surfaces.

If the stress state is projected onto a plane through O and normal to OH (i.e., look down from the top of the cylinder in a direction parallel to OH), Fig. 4-17 results. As mentioned earlier, this is called the π plane and has the equation

$$\sigma_1 + \sigma_2 + \sigma_3 = 0 \qquad (\text{i.e., } \sigma_m = 0)$$

and the original coordinate axes are 120° apart. Any hydrostatic component

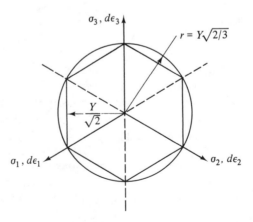

Figure 4-17 Tresca and von Mises yield surfaces projected onto the π plane.

is perpendicular to the π plane so it cannot induce plastic work since there is no tendency for the yield surface to expand. However, the deviatoric components, being perpendicular to the surface, would cause plastic work if they caused the yield surface to expand. Referring back to Fig. 4-14, maximum work occurs when the total strain vector ($d\epsilon_v$ in Fig. 4-16) is normal to the yield surface and in line with the deviatoric component of the total stress.

Now consider the meaning of the above comments in terms of the yield locus shown in Fig. 4-18. Note that although $\sigma_2 = 0$, $d\epsilon_2$ is not zero *in most*

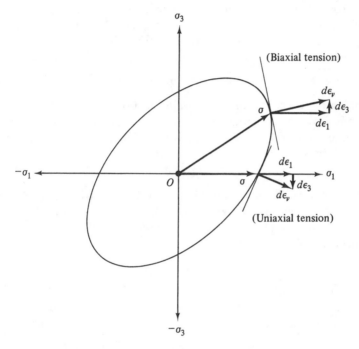

Figure 4-18 Illustration of the principle of normality as related to a yield locus.

cases. What is shown as $d\epsilon_v$ is really the projection of $d\epsilon_v$ onto the 1-3 plane; note that there is usually a component $d\epsilon_2$, which means that the tip of the total strain vector is tilted up from the plane of the page. However, this would not affect the normality of $d\epsilon_v$ as shown in Fig. 4-18. Note too that σ represents the vector sum of σ_1 and σ_3 and could again be broken down into another pair of components representing σ_m and the deviatoric stress.

One of the valuable aspects of normality is its use in the construction of yield surfaces and loci based upon experimental findings. Backofen [8] demonstrates this for a number of situations.

4-12 PLASTIC WORK

Consider a bar of length l_0 subjected to a tensile force F acting on an area $(w_0 t_0)$ with a resulting extension dl. The work done is $F\, dl$ and on a unit volume basis is:

$$dW_v = \frac{F\, dl}{w_0 t_0 l_0} = \frac{F}{w_0 t_0} \cdot \frac{dl}{l_0} = \sigma\, d\epsilon \qquad (4\text{-}30)$$

If a shear force caused deformation, a similar argument would show that $\tau\, d\gamma$ expresses the work per unit volume done by that force. These individual contributions could be summed up in a manner that produced Eq. (4-10) noting that the coefficient of one-half does not appear in Eq. (4-30). In terms of principal stresses:

$$dW_v = \sigma_1\, d\epsilon_1 + \sigma_2\, d\epsilon_2 + \sigma_3\, d\epsilon_3 \qquad (4\text{-}31)$$

In defining the effective stress and strain functions by Eqs. (4-8) and (4-26), no mention was made about plastic work, but one might now expect that such definitions could be used in this regard. This is indeed the case and the resulting expression is:

$$dW_v = \bar{\sigma}\, d\bar{\epsilon} \qquad (4\text{-}32)$$

Two examples will demonstrate this equivalence. Consider uniaxial tension where $\sigma_1 \neq 0$, $\sigma_2 = \sigma_3 = 0$ and $d\epsilon_2 = d\epsilon_3 = -\frac{1}{2}\, d\epsilon_1$ (since $d\epsilon_1 + d\epsilon_2 + d\epsilon_3 = 0$). Using Eq. (4-31) gives $dW_v = \sigma_1\, d\epsilon_1$ since the other terms vanish. The effective stress and strain functions show that

$$\bar{\sigma} = \sigma_1 \quad \text{and} \quad d\bar{\epsilon} = d\epsilon_1$$

so equivalence results.

Next consider pure shear where $\sigma_1 = -\sigma_3$, $\sigma_2 = 0$ and $d\epsilon_1 = -d\epsilon_3$, $d\epsilon_2 = 0$. From Eq. (4-31),

$$dW_v = \sigma_1\, d\epsilon_1 + (-\sigma_1)(-d\epsilon_1) = 2\sigma_1\, d\epsilon_1$$

The relations in terms of effective values show that

$$\bar{\sigma} = \sqrt{3}\,\sigma_1 \quad \text{while} \quad d\bar{\epsilon} = \frac{2}{\sqrt{3}}\, d\epsilon_1$$

Thus,

$$dW_v = (\sqrt{3}\,\sigma_1)\left(\frac{2}{\sqrt{3}}\, d\epsilon_1\right) = 2\sigma_1\, d\epsilon_1$$

as before.

Thus, besides finding use in regard to yielding predictions and the flow rules, the concepts of effective stress and strain provide a convenient way to calculate the work due to plastic deformation.

4-13 COMPARISON OF MOHR'S CIRCLES FOR STRESS AND PLASTIC STRAIN INCREMENTS

Since the hydrostatic component of the total stress causes no plastic flow, this implies that the value of σ_m on a circle plot of stresses should coincide with the origin of the incremental strain circle (i.e., $d\epsilon = 0$). Under uniaxial tension, $\sigma_m = \frac{1}{3}\sigma_1$, $\sigma_2 = \sigma_3 = 0$, while $d\epsilon_2 = d\epsilon_3 = -\frac{1}{2}d\epsilon_1$. Figure 4-19(a) shows the physical meaning that relates the stress circle to the scaled results in terms of strains; note the negative two-to-one correspondence of incre-

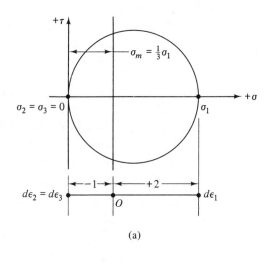

(a)

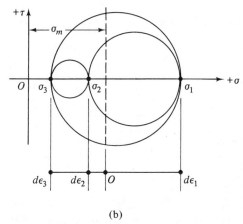

(b)

Figure 4-19 Relationship of Mohr's circle of stress and incremental plastic strains for uniaxial tension (a) and a general triaxial situation (b).

mental strains. Figure 4-19(b) shows a more general case where $\sum d\epsilon_i$ must equal zero.

It is also instructive to consider plane strain, where from Eq. (3-13):

$$\sigma_2 = \tfrac{1}{2}(\sigma_1 + \sigma_3) \qquad \text{since } v = \tfrac{1}{2} \text{ for plastic flow}$$

For this case, $\sigma_m = \sigma_2$ so the *zero* of the incremental strain circle coincides with σ_2 or $d\epsilon_2 = 0$. This result can be readily checked using Eq. (4-25).

It should be realized that changing σ_m by superimposing an additional uniform normal stress would displace the stress circle but would have no effect on the size of either set of circles. Thus, a knowledge of strain ratios does not *uniquely* define the current stresses.

REFERENCES

[1] H. Ford, *Advanced Mechanics of Materials* (New York: John Wiley & Sons, Inc., 1963), pp. 22–28.

[2] B. Paul, *Fracture, Vol. II* (New York: Academic Press, 1968, ed. H. Liebowitz), pp. 446–53.

[3] H. Tresca, *Comptes Rendus Acad. Sci. Paris*, 59 (1864), p. 754.

[4] R. von Mises, *Gött. Nach., math.-phys., Klasse* (1913), p. 582.

[5] W. Johnson and P. B. Mellor, *Engineering Plasticity* (London: Van Nostrand Reinhold Company, 1973), pp. 81–83.

[6] R. Hill, *Plasticity* (Oxford at the Clarendon Press, 1950), pp. 50–52.

[7] D. C. Drucker, *Proc. 1st U. S. National Congress of Applied Mechanics*, 1951, p. 487.

[8] W. A. Backofen, *Deformation Processing* (Reading, Mass.: Addison-Wesley Publishing Company, 1972), pp. 58–72.

PROBLEMS

4-1 Consider the stress states posed in Problem 1-5.
   (a) Determine the mean normal stress, σ_m, for each condition.
   (b) Find the three deviatoric stresses for the three principal directions.
   (c) What is the sum of the three deviatoric stresses in (b)?

4-2 Show that the "reduced stress invariant," J_2, equals

$$\tfrac{1}{6}[(\sigma_1 - \sigma_2)^2 + (\sigma_2 - \sigma_3)^2 + (\sigma_3 - \sigma_1)^2]$$

4-3 The tensile yield strength of a metal is given as Y ksi. If a specimen of this material were subjected to two normal compressive stresses of $-\tfrac{1}{4}Y$ and $-\tfrac{1}{2}Y$ acting along say x and y directions, what tensile stress applied in the z

direction would cause yielding according to:
(*a*) von Mises?
(*b*) Tresca?

4-4 A cube of metal, 1 in. on a side, is subjected to a compressive load of 50,000 lb at which point the metal yields. Assume that friction at the loading interface is negligible (i.e., no shear stresses). If an identical cube were first constrained by compressive loads of 20,000 and 30,000 lb on the other pairs of faces (again, friction is zero), what compressive load in the third coordinate direction would be necessary to cause yielding? Use the von Mises criterion.

4-5 A long, thin-walled tube, capped on the ends is made of a metal whose yield strength in uniaxial tension is 40 ksi. The tube is 60 in. long, has a wall thickness of 0.015 in. and a dimaeter 2 in. Under service conditions the tube experiences an axial tensile load of 1000 lbf, a torque of 1000 lb-in. and is to be pressurized internally. At what internal pressure is yielding predicted by:
(*a*) the Tresca criterion?
(*b*) the von Mises criterion?

4-6 A pressure vessel, in the form of a cylinder with hemispherical ends, has a radius of 2 ft and is to be made from a metal whose $k = 80$ ksi. The maximum internal pressure intended during use is 5 ksi. If no section of the vessel is to yield, what minimum wall thickness should be specified according to:
(*a*) Tresca?
(*b*) von Mises?

4-7 Consider the cases of (a) uniaxial tension, (b) pure shear, and (c) the triaxial case where $\sigma_1 > \sigma_2 > \sigma_3$ respectively. For each case compare the ratio of the effective stress, $\bar{\sigma}$ and the maximum shear stress, τ_{max}.

4-8 A thin-walled cylinder (diam. = 80 mm, wall thickness = 3.5 mm) just yields when a uniform axial stress of 200 MPa is applied. If an identical cylinder is loaded in bending to a maximum axial normal stress of 140 MPa, calculate the internal pressure required to cause yielding, using the Mises criterion.

4-9 A metal whose yield in pure shear is 15 ksi is formed into a thin-walled tube capped on the ends where: diam. = 6 in., wall thickness = 0.075 in., length = 20 in. The elastic modulus is 20×10^6 psi and $\nu = 0.25$. If the tube is pressurized to a level of 500 psi above that at which initial yielding occurred according to the von Mises criterion, determine the final length of the tube. *Note:* Ignore end effects, but do not ignore elastic effects.

4-10 Consider plane strain plastic deformation ($d\epsilon_2 = 0$). Using the von Mises criterion, show that $\sigma_1 - \sigma_3 = (2\bar{\sigma})/\sqrt{3}$.

4-11 A metal flows at a constant yield stress, i.e., $\bar{\sigma} = Y = $ constant. If this metal is deformed by uniaxial tension to a strain ϵ that does not induce necking, demonstrate that the plastic work per unit volume is $\bar{\sigma}\bar{\epsilon}$.

4-12 A material whose yield strength = 50 ksi is made in the form of a cube and subjected to a tensile stress, σ_1, along one axis and a stress $\sigma_3 = -\sigma_1/2$ along a second set of axes.

(a) Determine the ratio of the principal strain increment $d\epsilon_1/d\epsilon_2$.

(b) Using the von Mises criterion, determine the magnitude of τ_{max} at the onset of yielding.

(c) Repeat (b) where $\sigma_3 = +\frac{1}{2}\sigma_1$.

(d) Repeat (b) and (c) using the Tresca criterion.

4-13 A stress state is described in (ksi) by $\sigma_1 = 30$, $\sigma_2 = 15$, $\sigma_3 = 0$.

(a) Determine the strain ratio $d\epsilon_1/d\epsilon_3$.

(b) If a hydrostatic stress of 20 ksi is superimposed by fluid pressure upon the initial stress state, how does the ratio in (a) change? Explain.

4-14 A thin-walled cylinder (diam. $= 80$ mm, wall thickness $= 3.5$ mm) just yields when a uniform axial stress of 200 MPa is applied. If an identical cylinder is loaded in bending to a maximum axial normal stress of 140 MPa, calculate the internal pressure required to cause yielding, using the Tresca criterion. See Prob. 4-8.

4-15 A thin-walled cylinder just yields when the torsional shear stress reaches 43.8 ksi. If an identical cylinder is loaded to a torsional shear stress of 38.6 ksi, calculate the applied axial compressive stress necessary to just cause yielding. Assume the Tresca criterion holds.

4-16 A rigid-plastic cube is subjected to σ_1 on one pair of opposite faces, to $\sigma_2 = 0.2\sigma_1$ on a second pair, and to $\sigma_3 = -0.4\sigma_1$ on the third pair of faces. The stresses are gradually increased, maintaining the above ratios. Using the von Mises criterion for yielding (for $Y = 300$ MN/m^2), calculate the magnitudes of the principal stresses and the ratios of the three principal strain increments at the moment of yielding.

4-17 The von Mises criterion is to be used for yield predictions of a given metal. Under plane strain deformation where σ_3 is zero, make a sketch of the von Mises ellipse for the first quadrant in $\sigma_1 - \sigma_2$ stress space and

(a) Indicate how the strain increments $d\epsilon_1$ and $d\epsilon_2$ would look on the plot where normality prevails.

(b) Determine the plastic work increment in terms of the principal stresses and strain increments.

(c) Explain your answer in (b) with reference to the plot in (a).

4-18 A solid aluminum cylinder, 4 in. diameter by 6 in. length, is fitted inside a thin-walled tube of steel as shown. Through two hardened steel pistons, an axial load is to be transmitted to the aluminum and is indicated by forces F. Frictionless conditions may be assumed at all contact surfaces and when no force is applied, the outer surface of the aluminum is in intimate contact with the inner diameter of the steel tube. If $Y = 7$ ksi for the aluminum and 100 ksi for the steel, what force F must be applied to induce initial yielding of the tube? Use the von Mises criterion, and do not concern yourself with the elastic effects up to the onset of yielding!

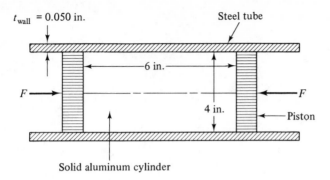

t_{wall} = 0.050 in.

Steel tube

6 in.

4 in.

F

F

Piston

Solid aluminum cylinder

5

OBSERVED MACROSCOPIC BEHAVIOR OF DUCTILE METALS

5-1 INTRODUCTION

The developments in previous chapters have been somewhat abstract, and it is appropriate to discuss certain aspects of mechanical behavior that provide a tie-in with the properties alluded to earlier. Although the bulk of this chapter could be considered as a review, the purpose here is to clarify the more important aspects of mechanical properties. Experience has indicated that many engineers have *never* been fully exposed to some of the findings determined from a simple tensile test. This is especially true in regard to the important topic of work hardening.

5-2 DEFORMATION OF DUCTILE METALS
BY UNIAXIAL TENSION

When metals are deformed plastically at temperatures below that at which recrystallization might occur, the terms *cold working, work hardening*, or *strain hardening* are used to describe this deformation. In addition to shape changes that result, certain properties of the original structure are altered as a consequence of plastic deformation. Use of the term work hardening implies in a qualitative sense that the metal becomes hardened and strengthened; it would, of course, be of greater advantage if a quantitative assessment of this change were made. Toward this end, plastic deformation by uniaxial tension provides a natural starting point for several reasons. First, the general mechanics involved in conducting such a test are simple to grasp and should be known by the reader. Second, the stress state involved is uncomplicated in comparison with other processes that induce plastic deformation. In addition, there exists a large body of numerical information in the literature that has accrued from tensile testing, and proper use of this information requires a thorough understanding of the test itself as well as the conditions to which such a test is restricted. Finally, no other deformation process lends itself so directly to a quantitative approach concerned with changes in mechanical properties caused by work hardening.

Although the words basic or fundamental are sometimes ascribed to property values determined by tensile testing, it must be appreciated that the conditions which prevail during a test play a significant role in regard to observed test results. In the general case, the following conditions are typical:

1. Strain rate is on the order of 10^{-3} per second.
2. Temperatures are on the order of 70 to 90°F (20 to 32°C).
3. Measurements are restricted to a gage section that is subjected to a state of uniaxial tensile stress.

5-3 LOAD VERSUS ELONGATION

The information obtained from a tensile test is the load or force required to cause a certain deformation or extension. Using these data, various stress-strain computations are then performed. Although the latter provide a more logical and useful set of findings, it is more basic to first discuss the load-elongation behavior. Regardless of the specimen shape used, test measurements are inevitably made on that section of the specimen throughout which the stress is assumed to be in a state of uniaxial tension. This section is designated to have some initial gage length and cross-sectional area which may be described as l_0 and A_0, respectively; thus, the initial or unloaded volume of metal of concern is $V_0 = A_0 l_0$, no attention being paid to that physical portion of the specimen which is outside of volume V_0.

As an ever-increasing axial load L is applied, the test section elongates and becomes smaller in area. Under some load L_1, the corresponding gage length and area would be l_1 and A_1, respectively. Up to some particular load level, the specimen deformation is elastic since removal of this load would permit the specimen to return to its original dimensions. This load is often referred to as the elastic limit. If such a load is exceeded, permanent deformation is said to result from plastic flow of the metal.

It is worth repeating at this point that during *elastic* deformation, the volume of the test section does not remain constant; rather the Poisson effect indicates that a volume change must occur unless Poisson's ratio is 0.5. For many metals, it is closer to 0.3. During *plastic* deformation however, volume constancy is noted for all practical purposes. In effect, volume constancy will always be implicit during plastic deformation; this requires that Poisson's ratio is 0.5. Volume changes arising from the initial elastic deformation are so small that they can be ignored. Example 4-4 demonstrated that the plastic component of deformation is often so much larger than its elastic counterpart, the assumption of volume constancy is quite reasonable. This greatly simplifies many analyses.

The above comments can be summarized by a general and qualitative plot of load-elongation behavior as shown in Fig. 5-1. Many important observations result from this plot, but first we note that the elastic behavior, from 0 to a is highly exaggerated to assist in the discussion that follows. Actual experiments with real metals would show that the elastic elongation, *om*, is so small in comparison with the plastic elongation, *mf*, that the line *oa* would be nearly vertical if all data were plotted to a common scale. With this in mind, we observe the following:

1. The portion of the curve shown as *oa* describes elastic behavior, elongation being directly proportional to the load for any point on that line; the maximum elastic elongation connected with this region is *om*.

2. The *yield load* L_y, which for our purposes is defined by point a, indicates the beginning of plastic deformation.

3. Continual increase of load above L_y follows the path *abc* at which point a maximum load L_u is reached. Note that as progress from a to b to c occurs, the initial gage length l_0 increases while the corresponding decrease in cross-sectional area is uniform throughout the full gage length.

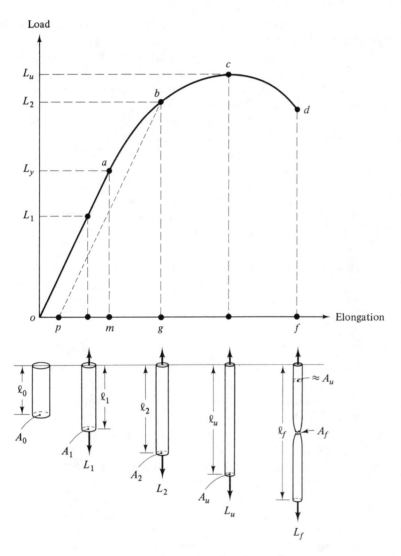

Figure 5-1 Load-extension plot of a tensile test.

4. Although the area is diminishing from *a* to *c*, a continual increase in applied load is required to cause further plastic deformation. This results because a ductile metal possesses the ability to harden or strengthen as it is deformed plastically. Up to the ultimate load, L_u, the rate of hardening offsets the diminishing area with the consequence that further plastic deformation demands a greater load.

5. When point *c* is reached, a condition of instability occurs and, at

some cross section of the test region, a localized constriction begins. This phenomenon is known as *necking* and the area reduction in this region is no longer offset by the added strengthening effect of plastic deformation. As a consequence, the load required to cause further elongation *in the vicinity of the neck* is lessened. What is often omitted in describing these events is that the usual tensile machine is designed to apply only enough force to the specimen to cause an additional increment of elongation. Thus, the behavior from c to d occurs because of the inherent ability of the testing machine to gradually reduce the applied load after a neck has started. If dead weights were used as a loading mechanism, any behavior described by Fig. 5-1 would terminate at point c because the coupling of load removal with the onset of instability would not exist; instantaneous fracture would occur at the load, L_u.

6. Once a neck has formed, practically all further plastic deformation from c to fracture at d occurs in the localized region of the neck. If one checked the area of the specimen away from the neck, as shown in Fig. 5-1, it would be *approximately* the same area that existed throughout the full test section when load L_u was first reached (i.e., A_u).

7. After point c is exceeded, the plastic deformation throughout the test section is no longer uniform as was the case from a to c.

8. The fracture area, A_f, is nominally taken to be the minimum area in the neck after the specimen has physically separated into two pieces.

9. If the specimen were originally loaded from o to b, and the load L_2 then fully removed, the unloading behavior could be adequately described by the line bp which is parallel to oa. Upon reloading, a linear relationship between load and elongation would then prevail up to point b; that is, the elastic region would be *extended* as compared with oa and, until the load L_2 was exceeded, additional plastic flow would not take place. Thus, the behavior of this specimen, which now contains some prior plastic deformation, would in effect follow the path $pbcd$.

5-4 *ENGINEERING OR NOMINAL STRESS AND STRAIN*

The limitation of a load-elongation plot is that no account is taken of the physical size of the test specimen. For example, if a metallic specimen having an initial area of 100 mm² could support a maximum load of 40 kN, there would be no fundamental significance to this *load* since a larger specimen of the same metal should be expected to carry a maximum load in excess of

40 kN. To avoid this shortcoming, it is sensible to convert the load-elongation data into some form of stress–strain data; most often the form of *engineering* stress and strain is used where the following definitions prevail.

$$\text{Engineering stress} = S \equiv \frac{L}{A_0} \tag{5-1}$$

$$\text{Engineering strain} = e = \frac{\Delta l}{l_0} = \frac{l - l_0}{l_0} \tag{5-2}$$

Since the load-elongation data in Fig. 5-1 are merely divided by the appropriate constants A_0 and l_0 to produce *nominal* stress-strain data, this amounts to a scale factor adjustment in both the ordinate and abscissa. Thus, a nominal stress-strain plot would be identical in form to that shown in Fig. 5-1. Throughout the elastic region *oa*, the linearity maintained between L and Δl would be duplicated in terms of stress and strain to produce

$$S = Ee \tag{5-3}$$

where E is the modulus of elasticity. The stress at point *a*, which has been assumed to indicate the onset of plastic deformation, is called the *yield strength* and is that level of stress at which plastic flow or yielding commences. It is defined as

$$S_y = Y \equiv \frac{L_y}{A_0} \tag{5-4}$$

To avoid an excessive use of subscripts, the symbol Y will be used to indicate the yield strength.

The stress level associated with the maximum load is referred to variously as tensile strength, ultimate tensile strength, and ultimate strength. *Tensile strength* will be used in this text and is defined as:

$$S_u \equiv \frac{L_u}{A_0} \tag{5-5}$$

Tensile strength signifies the maximum static *load* a particular section can support without fracturing. It is also sometimes used as a quality check.

Although the breaking strength is sometimes obtained by dividing the load at fracture, L_f, by the area A_0, it has no engineering significance and will be ignored on that basis. It is not at all representative of the stress state existing in the neck and any references to or use of such a value is meaningless.

Defining strain by Eq. (5-2), the maximum elastic strain associated with *oa* in Fig. 5-1 is the strain at yielding:

$$e_y = \frac{l_y - l_0}{l_0} \tag{5-6}$$

Consider now the strain induced by the application of load, L_2. According to the symbols used in Fig. 5-1:

$$e_2 = \frac{l_2 - l_0}{l_0} \tag{5-7}$$

which relates to the elongation *og*. This includes not only the strain caused by the plastic deformation from *a* to *b* but the total elastic deformation from *o* to *b*. It is the total strain induced due to load L_2.

The maximum *uniform* strain associated with the onset of necking is:

$$e_u = \frac{l_u - l_0}{l_0} \tag{5-8}$$

Any nominal strain based upon length measurements after the onset of necking provides nothing more than a useless average, since the strain throughout the gage section becomes more nonuniform as the severity of the neck increases.

5-5 MEASUREMENT OF YIELD STRESS

Reference to the yield stress has to now implied that there is a well-defined value of load which causes the onset of plastic flow. Point *a* in Fig. 5-1 is such a point. In reality, this is not true for most metals and has led to the necessity of defining such a point. Low-carbon steels containing no initial work hardening are often referred to when discussing a metal that has a pronounced yield point. This is somewhat misleading since such steels are an exception and do not typify the behavior of most other metals in regard to yielding. Reference to Fig. 5-2 will clarify these comments and it is noted again that the strain scale has been expanded to present this portion of the curve clearly.

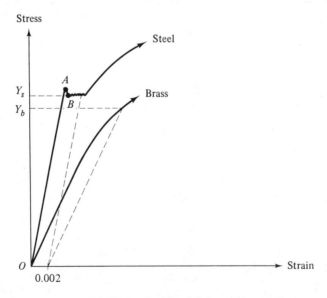

Figure 5-2 Offset method for defining yield strength.

Regarding the qualitative plot for a brass specimen, there is no specific load which causes additional strain at a constant stress; in effect there is no pronounced *yield point*. The most common practice regarding the yield strength of such metals is to employ the *offset method*. In Fig. 5-2 a 0.2 percent offset (i.e., a strain of 0.002) is utilized. This amounts to constructing a line that is parallel to the linear elastic portion of the curve and displaced by a strain of 0.002 as shown. Where this line intersects the stress-strain plot, that level of stress defines the yield strength for this particular offset; it is shown as Y_b. Two important observations should be noted. First, such a definition is arbitrary since other offset values from 0.1 percent to 0.5 percent are sometimes employed. Although for most metals the variation in yield strength that would occur for this offset range is relatively small percentage-wise, the exactness of such a value of yield strength is a bit nebulous. Secondly, the yield strength as defined in this manner is a stress that will produce a small amount of plastic deformation.

With low-carbon steel, a pronounced yield point is indicated, but even here the situation is not without question. Point A is usually called the *upper yield point*. The position of point A can be influenced by the rigidity of the testing machine as well as the rate of load application. Because of this, the lower yield point is generally considered as a more appropriate measure. If an offset were used, the intersection point, shown by Y_s, would lie on the flat portion of the stress-strain curve in the vicinity of point B.

Since the elastic modulus of brass is less than that for steel, the slopes of the lines up to yielding differ as shown. Therefore, the offset construction, although starting at 0.002 for both metals, must account for the variation in moduli.

5-6 INDICATIONS OF DUCTILITY

The extent to which a metal can be deformed plastically is referred to as the ductility of the metal. In an attempt to place a quantitative measure on this property, two parameters are most often employed; both are based upon measurements obtained after the test specimen has *fractured*.

Percent elongation is nothing more than the nominal strain at fracture multiplied by 100, or:

$$\text{Percent elongation} \equiv \left(\frac{l_f - l_0}{l_0}\right) 100 \qquad (5\text{-}9)$$

Percent reduction of area or *RA* is defined by

$$RA \equiv \left(\frac{A_0 - A_f}{A_0}\right) 100 \qquad (5\text{-}10)$$

In comparing the ductility of two metals, the one having a larger *RA* will usually possess a larger percent elongation, but there are exceptions. This can present a problem as to which of these measured quantities is more inherently descriptive of ductility. What can be affirmed is that percent elongation is more dependent upon the value of l_0 than is RA upon the value of A_0. Due to the localized strain concentration after necking has occurred, the smaller the initial gage length, the greater would be the percent elongation. This assumes, of course, that the neck lies within the initial length l_0. On the other hand, there is no inherent reason why RA should be greatly affected if various specimens of different A_0 values are employed. Although there is far more information in the literature giving values of percent elongation, RA is often more useful. Naturally, if necking were foreign, and fracture occurred at the maximum load, these two quantities would be directly related since the deformation would be uniform for the entire test. Examples 5-1 and 5-2 should make this clear.

Example 5-1.

A ductile metal undergoes tensile deformation until it fractures when the maximum load is reached. At that time the gage length was 60 mm while the area was 83.33 mm². Prior to loading, the gage length was 50 mm whereas the area was 100 mm². Determine the true strain at fracture using both length and area changes and explain why these results occur. Also, find the percent elongation and reduction of area at fracture.

Solution.

$$\epsilon_f = \ln\left(\frac{\ell_f}{\ell_0}\right) = \ln\left(\frac{60}{50}\right) = 0.182$$

$$\epsilon_f = \ln\left(\frac{A_0}{A_f}\right) = \ln\left(\frac{100}{83.33}\right) = 0.182$$

They are the same because the deformation was uniform up to the point of fracture, that is, necking did not occur since fracture occurred at the maximum load.

$$\text{Percent elongation} = \frac{60 - 50}{50}(100) = 20 \text{ percent}$$

$$\text{Percent reduction of area} = \frac{100 - 83.33}{100}(100) = 16.67 \text{ percent}$$

Assuming constancy of volume and uniform deformation:

$$V_0 = A_0 l_0 = A_f l_f = (0.833)A_0(1.2)l_0 = A_0 l_0$$

Example 5-2.

A tensile specimen of annealed 1020 steel is pulled to fracture. The initial gage section dimensions were $l_0 = 2$ in. and $d_0 = 0.505$ in. After reaching the maximum load, a neck forms and continues to get smaller until fracture occurs. The final gage length is 2.74 in. while the smallest diameter in the neck is 0.332 in. Repeat Example 5-1 using these test data.

Solution.

$$\epsilon_f = \ln\left(\frac{2.74}{2}\right) = 0.315$$

$$\epsilon_f = 2\ln\left(\frac{0.505}{0.332}\right) = 0.838 = \ln\left(\frac{A_0}{A_f}\right) = \ln\left(\frac{0.2}{0.087}\right)$$

In this case, the fracture strain cannot be determined using length values since the strain is very nonuniform due to necking, thus, the 0.315 value is meaningless. The value based upon the starting area and *minimum* area in the neck is at least representative of the *maximum* strain experienced by the specimen.

$$\text{Percent elongation} = \frac{2.74 - 2}{2}(100) = 37 \text{ percent}$$

$$\text{Percent reduction in area} = \frac{0.2 - 0.087}{0.2}(100) = 56.5 \text{ percent}$$

Since nonuniform deformation occurred, the above values cannot be used to show constancy of volume as in Example 5-1.

A common practice used to express the extent of induced plastic deformation is to introduce the term *percent cold-work*. This is defined for a particular reduction of area as follows:

$$\text{Reduction of area} = r = \frac{A_0 - A}{A_0} \qquad (5\text{-}11)$$

The percent cold-work is simply r multiplied by 100. Note that the maximum value that r can attain is governed by Eq. (5-10). The maximum uniform cold-work that can be imparted by uniaxial tensile deformation is determined from Eq. (5-11) where the area at ultimate load, A_u, is used as A.

5-7 TRUE STRESS AND TRUE STRAIN

For plasticity work, tensile data are most useful if the instantaneous stress and strain are associated with the current load level and total deformation that has occurred. The type of behavior shown by the path *abcd* in Fig. 5-1, even when converted to values of S and e, finds limited use in plasticity. Rather, it is more useful to define a *true* or *natural* stress and its strain counterpart. True or logarithmic strain has been discussed in Chap. 2 where it was defined as:

$$d\epsilon = \frac{dl}{l}$$

or after integration,

$$\epsilon = \ln\left(\frac{l}{l_0}\right) \qquad (5\text{-}12)$$

Because of volume constancy, equivalent expressions for true strain are:

$$\epsilon = \ell n \left(\frac{l}{l_0}\right) = \ell n \left(\frac{A_0}{A}\right) = \ell n \left(\frac{1}{1-r}\right) = 2 \ell n \left(\frac{D_0}{D}\right) \qquad (5\text{-}13)$$

True stress is defined as

$$\sigma \equiv \frac{L}{A} = \frac{\text{instantaneous load}}{\text{instantaneous area}} \qquad (5\text{-}14)$$

Before proceeding to use these new definitions, we illustrate the additive property of true strains, which were mentioned in Chap. 2.

Using Eq. (5-2) and assuming tensile loading, suppose $l_0 = 2$ in.; it is then increased in sequence by $\Delta l_1 = 0.2$, $\Delta l_2 = 0.2$, $\Delta l_3 = 0.2$:

$$e_1 = \frac{2.2 - 2}{2} = \frac{0.2}{2} = 0.1$$

$$e_2 = \frac{2.4 - 2.2}{2.2} = \frac{0.2}{2.2} = 0.091$$

$$e_3 = \frac{2.6 - 2.4}{2.4} = \frac{0.2}{2.4} = 0.083$$

$$e_t = e_1 + e_2 + e_3 = \underline{0.274}$$

If the full deformation is done in one step, $e_t = (2.6 - 2)/2 = 0.300 \neq 0.274$.

Now using Eq. (5-13) and the same sequence,

$$\epsilon_1 = \ell n \left(\frac{2.2}{2}\right) = \ell n \,(1.1) = 0.095$$

$$\epsilon_2 = \ell n \left(\frac{2.4}{2.2}\right) = \ell n \,(1.09) = 0.086$$

$$\epsilon_3 = \ell n \left(\frac{2.6}{2.4}\right) = \ell n \,(1.083) = 0.08$$

so

$$\epsilon_t = \epsilon_1 + \epsilon_2 + \epsilon_3 = 0.261$$

If the deformation is computed in a single step, $\epsilon_t = \ell n \,(2.6/2) = \ell n\, 1.3$ or $\epsilon_t = 0.261$ which is the same as the sum of the individual values just computed. It must be cautioned here, that if changes in loading directions are severe, or plastic strain reversals are encountered, the above procedure may not accurately describe the actual true strain induced.

5-8 WORK HARDENING BY UNIAXIAL TENSION

The engineering stress-strain curve and the corresponding true stress-strain curve for a general set of tensile test data are plotted in Fig. 5-3, where the differing definitions prevail. Although S and σ can be plotted to the same scale, the strain scale must be handled properly since Eqs. (5-2) and (5-12) can be used to show that:

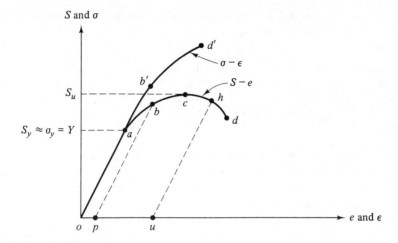

Figure 5-3 Comparison of engineering (nominal) and true (logarithmic) stress-strain curves. Up to point c, $\epsilon = \ell n\,(1 + e)$ and $\sigma = S(1 + e)$ are valid relations.

$$\epsilon = \ell n\,(1 + e) \qquad (5\text{-}15)^*$$

As the specimen is loaded from o to a, the decrease in area is wholly negligible so that $A_o \approx A_a$. Thus, the yield strength would be defined by:

$$Y = \frac{L_y}{A_a} \approx \frac{L_y}{A_o} \qquad \text{or} \qquad Y = S_y \approx \sigma_y \qquad (5\text{-}16)$$

In words, the yield stress is for all practical purposes a true stress as defined by Eq. (5-14) and there is no need for differentiating between the nominal stress at yield and the true stress at yield. They are equivalent and will be referred to hereafter only as the *yield strength*.

Example 5-3.

Consider a rigid perfectly plastic solid. Construct a qualitative load-extension plot for such a solid subjected to uniaxial tension and explain this result.

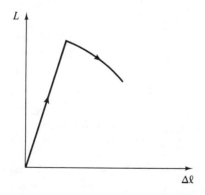

*When $e \leq 0.02$, $\epsilon \approx e$.

Solution.

As soon as initial yielding occurs, the load needed to cause further deformation begins to drop continuously. Referring to Fig. 4-1, once the yield stress is reached, further flow proceeds under a constant stress. Now for σ to remain constant as the specimen elongates and undergoes an area reduction, the load must decrease as deformation proceeds since $\sigma = L/A = $ constant.

Regarding the tensile strength as defined by Eq. (5-5), this is not a true stress as such since A_u is less than A_0. Suppose a second specimen, identical to the one that produced the plot *oabcd* in Fig. 5-3, was deformed plastically up to point b and then fully unloaded to point p. Obviously, this second piece has been work-hardened since it experienced plastic deformation from a to b. What is its *new* yield stress? Upon reloading, yielding would not occur until a load associated with point b would be reached (i.e., L_b where $L_b > L_a$) so the yield load of this cold-worked specimen is greater than the yield load observed initially. In addition, the new *original* area is that area associated with point p (i.e., A_p where $A_p < A_0$) thus, the new yield stress would be:

$$Y = \frac{L_b}{A_p} \approx \frac{L_b}{A_b}$$

as in Eq. (5-16). Not only is the yield strength of the cold-worked specimen increased because of a higher yield load, but also because a smaller *original area* prevails. What is most important is that the new yield stress is a true stress which might correspond to b' on the σ-ϵ plot, so this form of a stress-strain plot shows the change in yield stress as a function of induced strain. With many ductile metals *that contain no initial work hardening prior to plastic deformation*, the fully plastic portion of the σ-ϵ curve (from a to d') can be represented by an empirical expression of the form:

$$\sigma = K\epsilon^n \quad \text{or} \quad \bar{\sigma} = K\bar{\epsilon}^n \qquad (5\text{-}17)^*$$

The physical interpretation of Eq. (5-17) is of great importance. If a certain degree of cold-work is induced by uniaxial tension, this corresponds to a certain value of r from Eq. (5-11) which is equivalent to a certain value of induced strain, ϵ, from Eq. (5-13). Assuming K and n are known for this metal, the calculated value of σ for this corresponding ϵ is the new yield stress of the work-hardened material. Thus, Eq. (5-17) can be interpreted as an indication of yield stress as a function of cold-work; it could be expressed as:

$$\bar{\sigma} = K\bar{\epsilon}^n = Y \qquad (5\text{-}18)$$

This provides a method for determining the quantitative effect of cold-working on the resulting yield stress. With regard to the change in tensile strength due to cold-working, reference to Fig. 5-3 is necessary. The specimen

---

*That $\sigma_1 = \bar{\sigma}$ and $d\epsilon_1 = d\bar{\epsilon}$ (or $\epsilon_1 = \bar{\epsilon}$) during tensile deformation may be proven from the flow rules, Eq. (4-25).

loaded to point b, then unloaded to point p, contains a certain amount of cold-work. Just as its new yield stress was determined from:

$$Y = \frac{L_b}{A_p} \approx \frac{L_b}{A_b}$$

the new tensile strength can be found using Eq. (5-5) where

$$S_u = \frac{L_c}{A_p} \tag{5-19}$$

since it will reach the maximum load defined by point c and has an original area, *after the prior cold-work*, associated with point p. The tensile strength of this metal, when tested from a starting condition containing *no prior cold-work* would be:

$$S_u = \frac{L_c}{A_o} \tag{5-20}$$

Here it is necessary to distinguish symbolically between the value of S_u in Eqs. (5-19) and (5-20). Any reference to the tensile strength of a metal subjected to no prior cold-work *before* it is measured by a tensile test is S_u. Where the metal has been subjected to plastic deformation *before* being tested in tension, the tensile strength of this cold-worked metal will be designated as S_w. Using these symbols and combining Eqs. (5-19) and (5-20) gives:

$$S_w = \frac{S_u A_o}{A_p} = \frac{S_u}{1 - r} \tag{5-21}*$$

The magnitude of S_u is the value usually found in handbooks, and Eq. (5-21) permits predictions of the tensile strength of a cold-worked metal if the degree of cold-work is known.

At this point a *temporary restriction* must be placed upon the validity of Eqs. (5-18) and (5-21) since their development was based upon uniform deformation or straining. Therefore, for cold-working induced by uniaxial tensile deformation, they are valid to a point such as c in Fig. 5-3. Beyond this, the onset of necking leads to nonuniform deformation, and the methods used to calculate the change in these properties are questionable.

5-9 DETERMINATION OF THE WORK-HARDENING EQUATION

The curve $oab'd'$ in Fig. 5-3 portrays the true stress-strain plot on rectangular coordinates, and Eq. (5-17) has been introduced to describe the stress-strain behavior beyond point a. Rather than using curve-fitting for defining K and n in this equation, it is simpler to plot the data on logarithmic coordinates

*Apparently this was first proposed by Datsko [1].

initially since a power law equation plots as a straight line on such coordinates.

Figure 5-4 shows experimental values obtained for a commercially pure aluminum specimen which was *fully annealed* before being subjected to a tensile test. With logarithmic coordinates, there is no zero-zero starting point so the elastic region, depicted as Zone I, must start at some finite value. If the metal follows Hooke's law, the stress-strain relation in this zone obeys $\sigma = E\epsilon$ (or $S = Ee$) and the plotted points *must* form a 45° line to either axis. Although E is defined by the slope of the line when rectangular coordinates are used, the *slope* of this line on logarithmic coordinates bears no relation to the elastic modulus. The moduli of different materials are shown by the relative position of such lines with respect to each other. Their extrapolation to the intersection with unit strain defines the value of E; this is shown in Fig. 5-4. Materials such as plain low-carbon steel, having a pronounced yield point, exhibit the behavior shown in Fig. 5-5.

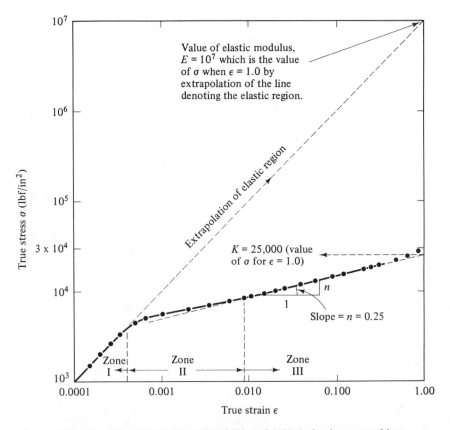

Figure 5-4 True stress–true strain behavior of 1100–0 aluminum tested in uniaxial tension [2]. Note the logarithmic coordinates.

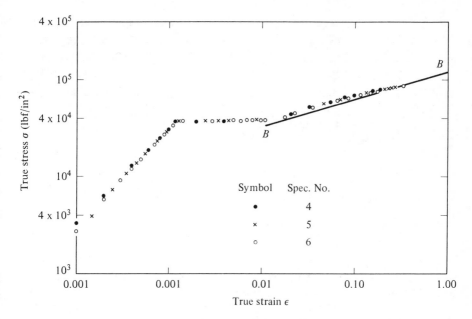

Figure 5-5 True stress–true strain behavior of annealed, low-carbon steel in uniaxial tension [3]. Note the logarithmic coordinates.

Zone II represents a transition from elastic to fully plastic behavior and is typical of most ductile metals. Once Zone III is reached, the metal may be viewed as being in a fully plastic condition and from this point up to necking, the measured test points fall on a straight line. This is true if the specimen had received no cold-work *prior to testing*. The slope of this line defines the strain-hardening exponent, *n*, and the intersection of this line with unit strain gives the stress value that defines the magnitude of *K*. Figure 5-4 shows these results. Thus, we have:

$$\sigma = E\epsilon^{1.0} \quad \text{(elastic-Zone I)} \quad (5\text{-}22a)$$

$$\sigma = K\epsilon^{n} \quad \text{(plastic-Zone III)} \quad (5\text{-}22b)$$

and the similarity of these relations might permit reference to *K* as a plastic modulus. Such terminology, however, is not used.

Since no simple equation describes the stress-strain behavior of real metals from the onset of loading to fracture, objections are sometimes raised to the use of Eq. (5-18) for describing strain-hardening behavior. Of course, other forms of strain-hardening relations must be subject to these same criticisms. It is essential, therefore, to realize the physical restrictions that must be considered when a power law form of strain hardening is utilized. These can be enumerated:

1. Such an equation is quite reliable when the induced strain is greater than 0.02 but less than the strain at which necking begins.

2. Use of this equation to predict the *initial* yield strength of the metal should be avoided. Instead, the offset method should be used.

3. Most metal-working operations impart strains far in excess of 0.02 (this is equivalent to about 2% cold-work!), and the exclusion of the elastic and transition strain regions leads to little error in this regard.

Since by definition:

$$\bar{\sigma} = \frac{L}{A} = K\bar{\epsilon}^n \quad \text{and} \quad d\bar{\epsilon} = \frac{dl}{l} = -\frac{dA}{A}, \tag{5-23}$$

at the instability point (maximum load) where $dL = 0$, it can be shown that $\bar{\epsilon}_u = n$. The derivation is posed in Prob. 5-1. Although this is mathematically correct, it is very difficult to measure the true strain at the exact maximum load because of the machine sensitivity that would be required; consequently, a plot such as Fig. 5-4 is the most reliable method for defining n. Often, however, a question can arise concerning the placement of the line that best describes Zone III behavior. We approach this problem as follows.

The true stress at ultimate load (note that it is dangerous to call this the true tensile strength as confusion might result) can be expressed as:

$$\bar{\sigma}_u = K\bar{\epsilon}_u^n = Kn^n \quad \text{since } \bar{\epsilon}_u = n \tag{5-24}$$

Using Eqs. (5-5), (5-14), and (5-24):

$$L_u = S_u A_o = \bar{\sigma}_u A_u = (Kn^n)A_u \tag{5-25}$$

Therefore,

$$S_u = (Kn^n)\frac{A_u}{A_o}$$

From Eq. (5-13),

$$\frac{A_u}{A_o} = e^{-\epsilon_u} \tag{5-26}$$

So,

$$S_u = K\left(\frac{n}{e}\right)^n \quad \text{since } n = \epsilon_u \tag{5-27}$$

where e is the base of natural logarithms in Eqs. (5-26) and (5-27).

Now once a straight line is fitted to test points, thereby defining K and n, S_u can be *calculated* using Eq. (5-27). If this varies from the *measured* value of tensile strength (say by $\pm 3\%$), it is probable that improper weight has been given to points in the Zone II region; a slight adjustment of the line would be needed. This technique provides a very useful *check* in assessing the proper values of K and n.

Example 5-4.

The work-hardening behavior of an initially annealed brass follows a power law expression of the form $\bar{\sigma} = 105{,}000\,\bar{\epsilon}^{0.5}$. If a piece of this metal were subjected to 30% cold-work in a uniform manner, what would be the expected yield strength and tensile strength of the plastically strained piece?

Solution.

From Eq. (5-13), the induced strain can be found since $r = 0.3$:

$$\bar{\epsilon} = \ell n \left(\frac{1}{1 - 0.3} \right) = 0.357.$$

The expected *yield* strength is determined from Eq. (5-18) as:

$$Y = \bar{\sigma} = K\bar{\epsilon}^n = 105{,}000(0.357)^{0.5} = 62{,}737 \text{ psi}$$

The tensile strength of this annealed brass is predicted from Eq. (5-27) to be:

$$S_u = 105{,}000\left(\frac{0.5}{e}\right)^{0.5} = 45{,}030 \text{ psi}$$

For this situation, the tensile strength of the cold-worked piece is found from Eq. (5-21) to be:

$$S_w = \frac{45{,}030}{1 - 0.3} = 64{,}330 \text{ psi}$$

It has been implied that the parameters K and n are material constants, and it is essential that this be understood thoroughly. One cannot, for example, assume that the magnitude of these parameters is fixed for a metal whose structure can be significantly altered by heat treatment. If a piece of SAE 1020 steel were fully annealed while a second piece had been austenitized and oil-quenched, different values of K and n would be found for these two specimens. Again, specimens of 2024 aluminum that were solution-treated and age-hardened would produce different values of K and n when compared with this same metal in an over-aged condition. In effect, each chemical composition and condition of microstructure must be viewed as a *different* metal as far as K and n are concerned. What must be realized, however, is that the value of the parameters should be determined using specimens that *contain no effect of work hardening prior to the tensile deformation itself.* After all, K and n are the very parameters used to define the work-hardening characteristics. Now, certain sources indicate that the strain-hardening parameters *vary* with the degree of prior cold-working. For example, a metal when fully annealed may produce a value of $K = 100{,}000$ psi and $n = 0.25$. If this metal were initially cold-worked 20% and then tested in tension, K might be shown as 115,000 psi and $n = 0.20$. There is a *fundamental flaw* in such an approach. The two values given for this metal in an initially annealed condition would be the correct parameters, and these *do not vary with prior cold-work.* If one pauses to consider Eqs. (5-16), (5-17), and (5-18), it should be obvious that the second specimen above (i.e., one initially cold-worked 20%) has been sub-jected to some initial increment of strain equivalent to 0.223 (where $r = 0.20$). Thus, the added effects of further plastic deformation can be easily, and *correctly*, handled by summing up the initial and additional plastic strain to give the total strain induced. This means that:

$$Y = K\bar{\epsilon}_t^n = 10^5(0.223 + \epsilon)^{0.25}$$

as the general strain-hardening equation for the second specimen. In this way the prior work-hardening effect is simply lumped together with any added strains and the *summation of strains* is raised to the exponent n. If this is understood, it is unnecessary to express a more general form of this equa-tion as:

$$Y = K(\bar{\epsilon}_o + \bar{\epsilon}_a)^n \tag{5-28}$$

where $\bar{\epsilon}_o$ represents the strain due to prior cold-working and $\bar{\epsilon}_a$ the additional strain due to subsequent cold-working. To illustrate this point more fully, a plot on logarithmic coordinates, as shown in Fig. 5-6, is helpful.

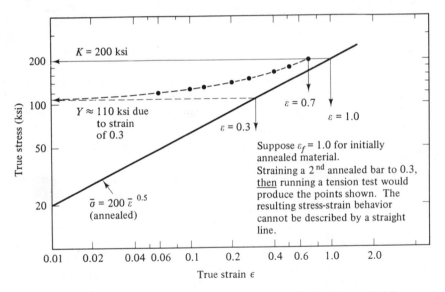

Figure 5-6 Comparison of tensile true stress–true strain data for a material initially annealed versus the same material containing work hardening prior to tensile testing.

5-10 BEHAVIOR AFTER NECKING

After Eq. (5-21), mention was made of a temporary restriction regarding Eqs. (5-18) and (5-21) if strains greater than $\bar{\epsilon}_u$ were induced. That is, how reliable are these relationships if the degree of cold-work imparted to a metal causes a strain in excess of that value which would lead to necking *under tensile deformation*?

First, consider tensile deformation itself. Once a neck forms, the stress state throughout the gage section is no longer uniaxial or uniform. As the smallest section in the neck (which has then been subjected to the greatest degree of cold-work) attempts to contract, it experiences constraint from the adjacent metal. Because of this, its radial reduction is hindered and the neck is subjected to a state of triaxial tension. In fact, the axial stress, due to the applied load, is no longer uniform across this section, but varies from a minimum at the outer surface to a maximum along the centerline. Needless to say, the situation is now far more complex than that which prevailed up to necking.

If the round test specimen were isotropic one could reasonably measure the minimum diameter of the neck and use this to compute the apparent maximum true stress and true strain at a particular instant. However, since the axial stress is not uniform across this minimum section, it should not be expected to agree with the uniform behavior that prevailed up to necking; it does not. In fact, the stress-strain points so obtained after necking occurs begin to fall *above* the line described by Eq. (5-18). This is demonstrated in Fig. 5-4. Correction factors have been developed to compensate for this condition of triaxiality and, if properly determined, tend to cause these points to fall on or very near the line described by Eq. (5-18) [4, 5]. This procedure compensates for the effects of necking, but it is rather tedious. Physically, however, it implies that the strain-hardening equation could be extrapolated and employed for strains in excess of $\bar{\epsilon}_u$ in tension. An apparently obvious solution would be to conduct tests in uniaxial compression. Unfortunately, *homogeneous* compression is difficult to achieve because of the influence of friction between the specimen ends and the compression loading surfaces. This leads to barrelling of the specimen as the applied load is increased; Fig. 5-7 illustrates this. As load is applied to a right circular specimen of original height and diameter h_o and d_o, the height becomes smaller and the area larger. At any particular load, the uniaxial true stress and strain would come from:

$$\sigma = \frac{L}{A} \quad \text{and} \quad \epsilon = \ell n \left(\frac{h_o}{h}\right)$$

both of these quantities being compressive as to sign. Because of the influence of friction at the workpiece-loading surface interface, the ends of the specimen are restricted from free radial expansion. As a consequence, wedge-shaped zones (shaded regions of Fig. 5-7) of relatively undeformed metal form at each end of the specimen and, in time, the nonuniform effect of barrelling becomes pronounced. To compute *uniform* stresses and strains under these conditions is not possible.

If the effects of friction can be minimized, then the existence of uniformity can be extended to greater deformation. It appears that the method of Cook

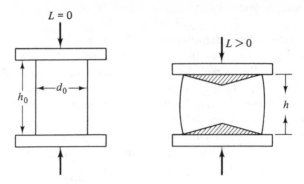

Figure 5-7 Direct compression showing formation of dead metal zones and barrelling.

and Larke [6] as later modified by Watts and Ford [7] provides the best means for obtaining true stress-strain compression data; it is, however, more time-consuming than is the tensile test. In general, the method utilizes cylinders of equal diameters but varying heights such that the d_o/h_o ratio varies from 0.5 to 3.0.* The modified approach is to start with well-lubricated ends and apply a particular load to a specimen, remove the load, and measure the new height. Upon relubrication, the specimen is subjected to an increased load, unloaded, measured, etc. This same approach is followed with each specimen where the particular *load* levels are exactly duplicated. Figure 5-8 shows how the results are plotted. For the same load, the actual reduction in height is

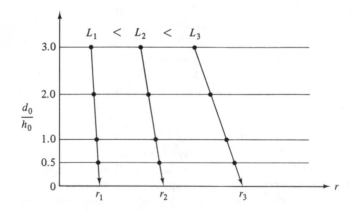

Figure 5-8 Watts and Ford [7] method for handling compressive loading to eliminate end effects.

*This must be limited to avoid buckling.

plotted against the diameter to height ratio for each cylinder. A line drawn through the points is extended to intersect a d/h ratio of zero; this would be equivalent to a specimen of infinite initial height, where the *end effects of friction* would be restricted to a small region of the full test height. In this way, it is theorized that the major portion of the specimen would deform uniformly.

With the value of r determined for a zero d/h ratio, the uniform true strain can be calculated using Eq. (5-13). As the original cross sectional area is known, and assuming the volume remains constant, the theoretical uniform change in area can be computed for a given load (say L_1 above). The true stress follows by dividing L_1 by this computed area; thus, a uniform true stress-strain combination is obtained. With a number of such points, the correct curve in compression can be plotted out to strains well in excess of the instability strain, $\bar{\epsilon}_u$, found in tension.

Examples exist in the literature [8] to illustrate that for many metals the plots of tensile and compressive values of true stress-strain are practically equivalent if sufficient care is exercised in obtaining the data. The important conclusion to be drawn is that the strain-hardening equation obtained from uniaxial tension can be employed at strains in excess of the necking tensile strain with some degree of confidence. Therefore, the tentative restriction placed upon Eq. (5-18) can be lifted.

Regarding the limitation of Eq. (5-21), it is best to discuss this on a physical basis. Referring to Fig. 5-3, suppose a specimen were loaded past the onset of necking, say to point h, at which time the load was removed. When the load is reapplied, further plastic flow would not occur until L_h was reached; this would also be the *maximum* load this specimen could support since the loading path would follow uhd. Obviously, the specimen is highly nonuniform at point h, but if it were possible to uniformly cold-work this metal *by some means other than tensile deformation* such that its entire structure were equivalent to that existing in the neck at point h, then its yield load and maximum load would be identical for all practical purposes. It follows, therefore, that:

$$S_w = Y = K\bar{\epsilon}^n \qquad \text{where } \bar{\epsilon} \geqq n \qquad (5\text{-}29)\dagger$$

Example 5-5.

If the material specified in Example 5-4 had been uniformly cold-worked to a 60% reduction in area, find the expected yield and tensile strengths of the strained piece. Also, draw the load-extension curve this piece would display when it was tested under tensile loading.

---

†On physical grounds, Y can never exceed S_w, but the two are practically equivalent for highly cold-worked metals.

Solution.

$$\bar{\epsilon} = \ell n \, (1/(1 - 0.6)) = 0.916$$

$$Y = 105,000 \, (0.916)^{0.5} = 100,500 \text{ psi}$$

Since the induced strain, 0.916, is greater than the strain-hardening exponent, 0.5, the discussion leading to Eq. (5-29) indicates that the tensile strength of the strained piece would approximate the calculated value of Y. The load-extension curve is as follows. Necking and yielding would occur simultaneously for all practical purposes and the load at which this takes place would also be the largest load observed. Therefore,

$$Y \approx S_w$$

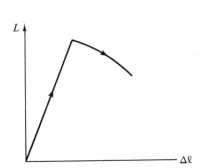

Two other experimental tests described below, are used to determine stress strain behavior to strains far in excess of the instability tensile strain. Because they require specialized equipment, they have not been used as extensively as the tensile test or direct compression test. Yet, both are informative.

5-11 BALANCED BIAXIAL TENSION (BULGE TEST)

In this test a disc of thin sheet is clamped around its periphery and fluid pressure is then applied on one side as shown in Fig. 5-9. As the sheet bulges, the tensile membrane stresses are

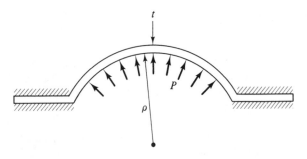

Figure 5-9 Thin sheet subjected to a bulge test.

$$\sigma_\theta = \sigma_\phi = \frac{Pp}{2t}$$

where P, p, and t are the instantaneous pressure, the radius of curvature, and the sheet thickness.

Now the thickness strain is $\epsilon_t = \ell n \, (t/t_o)$, t_o being the original thickness. From volume constancy, $\epsilon_t + \epsilon_\theta + \epsilon_\phi = 0$, and due to symmetry, $\epsilon_\theta = \epsilon_\phi$ (tensile strains). Thus,

$$\epsilon_t = -2\epsilon_\theta = -2\epsilon_\phi$$

The design of such a device requires that a relatively spherical shape occurs at the top of the dome and the measurements made directly are used to calculate σ_θ (or σ_ϕ) and the corresponding strain. The thickness strain (ϵ_t) which is compressive is simply twice the magnitude of ϵ_θ.

Consider a small stress element, shown in Fig. 5-10(a), where $\sigma_\theta = \sigma_\phi$. A superposition of a hydrostatic compressive stress (equal in *magnitude* to σ_θ), shown in Fig. 5-10(b), would give

$$\sigma_t = -\sigma_\theta$$

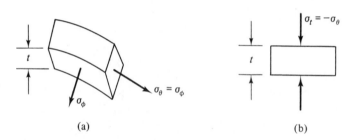

(a)   (b)

Figure 5-10 Equivalent stress states related to a bulge test.

Thus, the tensile loading is equivalent, stress-wise, to direct compressive loading and the resulting σ_t-ϵ_t plot depicts the through-thickness compressive stress-strain behavior. For reasonably isotropic materials these results coincide quite closely with those from a tension test, and the bulge test can be carried out to strains far in excess of the tensile strain at necking. Johnson and Mellor [9] give such results as well as additional references connected with this test method.

5-12 PLANE STRAIN COMPRESSION

This is another useful test which requires certain dimensional requirements of the testpiece. Nadai [10] first suggested this technique while Ford [11] appears to have been the first to actually use it. It is generally used on thin materials where, as shown in Fig. 5-11,

$$\frac{w}{b} > 6 \quad \text{and} \quad \frac{b}{t} \quad \text{from 2 to 4}$$

As the dies induce indentation, the unstressed metal outside the dies prevents any displacement or strain in the w dimension (or 2 direction). Incremental loading, usually in strain increments from $\frac{1}{2}$ to 2%, is used. The die load divided by die area gives the stress σ_1 while the change in thickness t leads to ϵ_1. Because of plane strain conditions, these results do not duplicate the

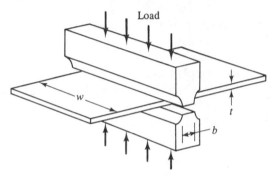

Figure 5-11 Details of a plane strain compression test.

σ-ϵ behavior from uniaxial tension or compression tests. If all such data are converted to effective stress and strain, the material is reasonably *isotropic* and care in experimentation is exercised, the stress-strain behavior due to uniaxial tension, direct compression, bulge testing, and plane strain compression can be adequately described by a single effective stress-strain function. This takes the form $\bar{\sigma} = K\bar{\epsilon}^n$ and because tensile testing is the simplest experiment to conduct, it provides the most practical way to define K and n. Unfortunately, other deformation modes do not induce strains of as uniform a nature as those from these tests. Because of this, the relation $\bar{\sigma} = K\bar{\epsilon}^n$ cannot be viewed as a universal law.

Example 5-6.

Under plane strain compression, determine the relationship between the applied compressive stress and the effective stress. Repeat this development for the effective strain and the strain in the direction of loading. Use the von Mises criterion for these purposes.

Solution.

Refer to Fig. 5-11 where the loading direction is taken as 1, the direction parallel to w as 2 and the direction parallel to b as 3.

With the condition of plane strain,

$$\sigma_2 = \tfrac{1}{2}(\sigma_1 + \sigma_3) \quad \text{and} \quad d\epsilon_2 = 0$$

(that is, no strain beneath the indenter in a direction parallel to w).

Also since there is no force opposing motion in the 3 direction,

$$\sigma_3 = 0$$

From these observations,

$$\sigma_2 = \tfrac{1}{2}\sigma_1 \quad \text{and} \quad d\epsilon_3 = -d\epsilon_1$$

Using Eq. (4-8):

$$\bar{\sigma} = \frac{1}{\sqrt{2}}\left[\left(\sigma_1 - \frac{\sigma_1}{2}\right)^2 + \left(\frac{\sigma_1}{2}\right)^2 + (\sigma_1)^2\right]^{1/2} = \frac{\sqrt{3}}{2}\sigma_1$$

Using Eq. (4-26):

$$d\bar{\epsilon} = \frac{\sqrt{2}}{3}[d\epsilon_1 - d\epsilon_2)^2 + (d\epsilon_2 - d\epsilon_3)^2 + (d\epsilon_3 - d\epsilon_1)^2]^{1/2} = \frac{2}{\sqrt{3}}d\epsilon_1$$

5-13 DUCTILITY OF PRIOR WORKED METALS

To this point, the quantitative effect of strain hardening on the yield and tensile strengths of ductile metals has been discussed. Of equal importance, is the alteration in ductility. The subject of fracture is exceedingly complex and the discussion that follows should certainly not be viewed as an attempt to predict fracture under complex loading systems. Instead, it is intended to give some feel to the observation that as cold-working increases strength, it lowers ductility. In essence, as the ability of a metal to flow plastically is utilized to strengthen a metal, the potential that remains for further plastic flow is diminished.

Suppose a tensile test is performed with a metal that has not been cold-worked prior to this test, and an RA of 60 percent results. Since this is based upon the smallest section in the neck of the specimen after fracture, the maximum strain experienced by this metal during this test is:

$$\bar{\epsilon}_f = \ell n\left(\frac{1}{1 - 0.6}\right) = 0.916 \qquad \left(r_{max} = \frac{RA}{100}\right) \tag{5-30}$$

A second specimen, identical in all aspects to the first one, might be cold-worked in tension, say to 10 percent reduction in area (i.e., $r = 0.1$), and a computation of the expected yield and tensile strengths of this cold-worked specimen could be made using the procedures in Example 5-4. However, interest might also center on a prediction of the new value of RA this *prior worked* specimen would exhibit. A cursory look might lead to the conclusion that an answer of 50 percent (i.e., $60\% - 10\%$) is correct. This is not the case since reduction in area is not additive in this sense. Using the concept of true strain, the answer becomes more apparent; with Eq. (5-30), the maximum strain this metal would experience in tension is 0.916. Due to the 10 percent cold-work initially imparted, the metal has experienced an initial strain of:

$$\epsilon = \ell n \left(\frac{1}{1 - 0.1}\right) = 0.105$$

and since true strains are additive, an additional strain of $0.916 - 0.105$ or 0.811 should cause fracture. Thus, the *RA* value for this prior worked specimen is computed from:

$$\epsilon = \ell n \left(\frac{1}{1 - r}\right) \qquad \text{or} \qquad \frac{1}{1 - r} = \exp(\epsilon)$$

The additional strain to cause fracture is 0.811 so that:

$$\exp(0.811) = \frac{1}{1 - r} \qquad \text{and} \qquad r = 0.555$$

or

$$RA = 55.5\%$$

Simple laboratory experiments will verify this concept.

Example 5-7.

Consider a ductile metal containing no prior effects of work hardening. Under a particular mode of deformation the piece fractures after being cold-worked 70 percent.

If another piece of the same annealed metal were first cold-worked 40 percent then machined into a tensile specimen, what reduction of area at fracture would be expected when the tensile test was completed?

Solution.

At first glance it might seem that the added cold-work to cause tensile fracture would be:

$$(70 - 40) = 30\% \qquad \text{or} \qquad r = 0.3$$

That this is not correct can be best shown by working with true strains, assuming that homogeneous deformation occurs regardless of the method of loading.

Using the maximum cold-work as 70%, the maximum strain at fracture is:

$$\epsilon_f = \ell n \left(\frac{1}{1 - r}\right) = \ell n \left(\frac{1}{1 - 0.7}\right) = 1.20$$

The strain induced initially in the second piece is:

$$\epsilon_i = \ell n \left(\frac{1}{1 - 0.4}\right) = 0.51$$

Thus the additional strain that would be expected at fracture is:

$$\epsilon = \epsilon_f - \epsilon_i = 1.20 - 0.51 = 0.69$$

To induce the strain of 0.69 requires a reduction of area of:

$$0.69 = \ell n \left(\frac{1}{1 - r}\right) \qquad \text{or} \qquad e^{0.69} = \frac{1}{1 - r}$$

so

$$r = \frac{0.994}{1.994} = 0.498$$

which does not agree with the value proposed earlier.

Example 5-8.

From Eq. (4-32), the plastic work per unit volume is equated to the area under the true stress-strain curve. A round bar of an annealed metal, whose strain-hardening behavior is expressed by $\bar{\sigma} = 200,000\bar{\epsilon}^{0.5}$, is originally 25 mm in diameter. By a series of two consecutive reductions with cylindrical dies, the initial diameter is reduced to 20 and 15 mm respectively as indicated in the sketch. Determine the plastic work per unit volume for each reduction and relate these findings to a graphical plot of the true stress–true strain curve.

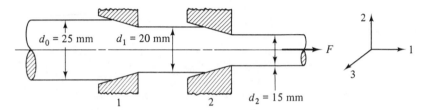

Solution.

$dW_v = \bar{\sigma}\, d\bar{\epsilon}$ gives the general relation to be used. For Step 1, due to axisymmetry, $\epsilon_2 = \epsilon_3$ so $\bar{\epsilon} = \epsilon_1$ since the strains are related as in uniaxial tension. If this is not obvious, Eq. (4-26) may be used to show that $\bar{\epsilon} = \epsilon_1$. $\bar{\epsilon} = \bar{\epsilon}_1 = 2\,\ell n\,(25/20) = 0.446$ which is the effective strain induced during the first reduction.

The total work per unit volume in Step 1 comes from

$$\int_0^{W_v} dW_v = \int_0^{\bar{\epsilon}_1} \sigma\, d\bar{\epsilon} = \int_0^{\bar{\epsilon}_1} 200,000\bar{\epsilon}^{-0.5}\, d\bar{\epsilon} = \frac{K\bar{\epsilon}^{(n+1)}}{n+1}\Big]_0^{\bar{\epsilon}_1}$$

$$W_{v_1} = \frac{200,000\bar{\epsilon}^{1.5}}{1.5}\Big]_0^{0.446} = 39,714\ \frac{\text{in.-lb}}{\text{in}^3}$$

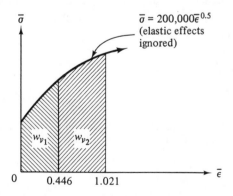

For Step 2, $\bar{\epsilon} = 2\,\ell n\,(20/15) = 0.575$ which is added to the initial strain of 0.446. Thus:

$$\int_{0.446}^{1.021} \bar{\sigma}\,d\bar{\epsilon}$$

is

$$W_{v_2} = \frac{200{,}000\bar{\epsilon}^{1.5}}{1.5}\Big]_{0.446}^{1.021}$$

$$W_{v_2} = 97{,}842\,\frac{\text{in.-lb}}{\text{in}^3}$$

Note that the *total* W_v for this entire process is $W_{v_1} + W_{v_2} = 137{,}556$ in.-lb/in³ which would also be found if the overall reduction had been completed in one step:

$$\bar{\epsilon} = 2\,\ell n\left(\frac{25}{15}\right) = 1.021 \text{ and } W_v = \int_{0}^{1.021} \bar{\sigma}\,d\bar{\epsilon} = 137{,}556 \text{ in.-lb/in}^3$$

5-14 SUMMARY

The various relationships developed in this chapter are based upon *homogeneous* plastic deformation. Except for the tensile test itself, and even here only to the onset of necking, no other deformation process causes so uniform a deformation; many cause deformation that is exceedingly nonhomogeneous. Thus, the expressions which predict the effect of cold-working on the changes in yield strength, tensile strength, and ductility cannot be considered as providing exact answers. Instead, they provide a first approximation. Often, with the use of necessary correction factors they can be employed to provide reasonable predictions of property changes of metals deformed by other loading modes. Of course, if such expressions were developed using other modes of loading (e.g., cold extrusion), a similar restriction would be required.

There is little question that an engineering approach to plastic deformation should make some attempt to include the effects of work-hardening. Until a better one is developed, the power law form appears to be as sensible as any other and will receive major use in this book. Note that in Examples 5-4, 5-5, and 5-8 and in Probs. 5-3, 5-9, 5-10, and 5-11, K has units of psi. Although K always has units of stress, the numerical value depends upon the system of units used.

REFERENCES

[1] J. DATSKO, *Material Properties and Manufacturing Processes* (New York, N.Y.: John Wiley & Sons, Inc., 1966), pp. 27–30.

[2] R. M. CADDELL and R. SOWERBY, "Strain Hardening: An Introduction to a Fundamental Experiment," *Bull. Mech. Engr. Educ.*, **8** (1969), pp. 31–43.

[3] R. M. CADDELL and J. H. LAMBLE, "Some Observations on the Tensile Testing of Ductile Metals," *Bull. Mech. Engr. Educ.*, **8** (1968), pp. 227–38.

[4] P. W. BRIDGMAN, "The Stress Distribution at the Neck of a Tension Specimen," *Trans. Am. Soc. Metals*, **32** (1944), pp. 553–74.

[5] N. N. DAVIDENKOV and N. I. SPIRIDONOVA, "Analysis of the State of Stress in the Neck of a Tension Test Specimen," *Proc. A.S.T.M.*, **46** (1946), pp. 1147–58.

[6] M. COOK and E. C. LARKE, "Resistance of Copper and Copper Alloys to Homogeneous Deformation in Compression," *J. Inst. Met.*, **71** (1945), pp. 371–90.

[7] A. B. WATTS and H. FORD, "On the Basic Yield Stress Curve for a Metal," *Proc. Inst. Mech. Engrs.*, **169** (1955), pp. 1141–49.

[8] E. G. THOMSEN, C. T. YANG, and S. KOBAYASHI, *Mechanics of Plastic Deformation in Metal Processing* (New York: The Macmillan Company, 1965), pp. 106–11.

[9] W. JOHNSON and P. B. MELLOR, *Engineering Plasticity* (London: Van Nostrand Reinhold Company, 1973), pp. 103–9.

[10] A. NADAI, *Plasticity* (New York: McGraw-Hill Book Company, 1931).

[11] H. FORD, "Researches into the Deformation of Metals by Cold Rolling," *Proc. Inst. Mech. Engrs.*, **159** (1948), pp. 115–43.

PROBLEMS

5-1 During a tensile test on a ductile metal that necks, the rate of change in load increase, dL, indicates a continual decrease until the maximum load is reached; at that time, $dL = 0$. This also corresponds to the instant when the true strain at maximum load, $\bar{\epsilon}_u$, occurs. Show that $\bar{\epsilon}_u = n$.

5-2 Show that $\epsilon = \ell n\,(1 + e)$ and $\sigma = S(1 + e)$. Discuss any limitations regarding the practical use of these relationships.

5-3 The strain-hardening behavior of a metal is given by $\bar{\sigma} = 100,000\bar{\epsilon}^{0.2}$.

    (a) If a piece of initially annealed material were uniformly cold-worked 10 percent, what would be the yield strength and tensile strength of this cold-worked material?

    (b) Repeat (a) if the material had been cold-worked 40 percent instead of 10 percent.

5-4 A specimen of original diameter of 0.357 in. and initial gage length of 2 in. is subjected to a standard tensile test. The following results are noted:

    (a) Yield load = 2000 lbf.

    (b) Fracture diameter = 0.270 in.

(c) Diameter at ultimate load = 0.310 in.
(d) Young's modulus = 25×10^6 psi.
(e) The strain-hardening exponent for this material is known to be 0.5.
(f) $\bar{\sigma} = K\bar{\epsilon}^n$
When this test is completed, you are told the specimen had been cold-worked some amount *before* the test which gave the above information was conducted!
(a) What is the yield strength of the test specimen?
(b) Knowing that $n = 0.5$, how much strain had been induced in the specimen by cold-work prior to the tensile test?
(c) What maximum load would be observed during the tensile test?
(d) What strain at fracture would be measured for the test specimen?

5-5 An annealed tensile specimen of alpha brass, with $A_0 = 0.2$ in² carries a maximum load of 12,000 lb and the initial area has been reduced 40 percent at that instant. If a second annealed specimen of identical starting dimensions were pulled until the induced true strain was half the magnitude of the strain-hardening exponent, what load in pounds would be required to reach this condition?

5-6 A tensile specimen of an annealed metal has a starting gage section of 0.505 in. diameter and 2 in. gage length. Test results indicate:
(a) Fracture diameter = 0.325 in.
(b) Yield load = 5000 lbf.
(c) Elastic modulus = 15×10^6 psi.
(d) Gage length at maximum load is 3 in.
(e) The maximum tensile load was 10,000 lbf.

1. What load would be expected at fracture for the above test?
2. When a true strain of 0.2 was induced during the above test, what would be the load in pounds on the specimen at that instant?
3. If a piece of this annealed metal were cold-worked 40 percent by rolling, what tensile yield strength would be expected?

5-7 It has been shown that the onset of necking, or tensile instability, for a metal pulled in uniaxial tension and having its strain-hardening behavior described by $\bar{\sigma} = k\bar{\epsilon}^n$ occurs when a strain of $\bar{\epsilon} = n$ has been reached. For metals whose strain-hardening behavior is described by the following forms, determine the value of strain at the onset of instability:
(a) $\bar{\sigma} = A(B + \bar{\epsilon})^n$ where B is a constant.
(b) $\bar{\sigma} = Ae^n$ where e is the nominal or engineering strain and $\epsilon = \ln(1 + e)$.

5-8 For a metal that strain hardens according to $\bar{\sigma} = K\bar{\epsilon}^n$ show that if it is strained to some level, say ϵ_1 corresponding to a stress σ_1, that the plastic work/unit volume is $\sigma_1\epsilon_1/(1 + n)$.

5-9 A material deforms plastically according to $\bar{\sigma} = 150,000(\bar{\epsilon})^{0.3}$
(a) Calculate the *force* necessary to *extrude* the solid from a 6-in. diameter bar to a 1-in. diameter bar.

(*b*) Calculate the *force* necessary to *draw* the solid from a 6-in. to a 1-in. diameter (drawing is pulling).

(*c*) Is (b) sensible? Prove your answer!

5-10 A thin-walled tube of annealed plain low-carbon steel has a strain-hardening relationship of $\bar{\sigma} = 10^5(\bar{\epsilon})^{0.22}$. Suppose it is placed over a mandrel that just fits the inner diameter of the tube and the tube is then work-hardened by alternately stretching and compressing. The sequence consists of a 5 percent stretch, then compression to its initial length per cycle. If the strain induced is homogeneous, how many cycles are required to produce a yield strength of 70 ksi?

5-11 A slab of 1100–0 aluminum, 1 by 10 by 50 in., is to be stretched to a 57-in. length, the 10-in. width being held constant. What is the maximum force needed to complete this operation if $\bar{\sigma} = 26,000(\bar{\epsilon})^{0.20}$?

5-12 Estimate the temperature rise in the neck of a tensile bar of a high-strength steel (yield stress $\approx$ 200,000 psi) necked to a reduction of area of 50 percent. Assume adiabatic conditions. Comment on the practicality of your answer.

5-13 Based upon the effective stress and strain functions given by Eqs. (4-8) and (4-26), compare the effective stress–effective strain behavior for each of the following methods of loading if uniform straining is involved:

(*a*) Uniaxial tension.

(*b*) Uniaxial direct compression.

(*c*) Plane strain compression.

(*d*) Plane strain tension.

(*e*) Balanced biaxial tension as in the bulge test.

5-14 A certain annealed metal is pulled to fracture and the maximum tensile strain is found to be 0.9 (i.e., $\epsilon_f = 0.9$). Suppose a piece of this annealed metal, initially 0.250 in. thick by 4 in. wide is cold-rolled to a thickness of 0.200 in. If you were told to cold-roll this sheet to fracture and were unaware that it had earlier been reduced in thickness to the 0.200 in. value, what percent reduction in area would you probably observe? Assume the 4 in. dimension remains constant.

6

VISCOELASTICITY

6-1 INTRODUCTION

Chapters 3 through 5 were concerned with macroscopic elastic and plastic deformation and strain was considered as a function of stress only. Although time was used as a third axis in the plots that described the behavior of appropriate models, the resulting deformations were time-independent. Elastic strains occurred instantaneously for any applied stress and were fully recoverable upon removal of load. How fast a stress might be applied, how long it was maintained and how quickly it was removed had no influence on the deformation behavior. To cause plastic deformation, a certain critical *level* of stress had to be exceeded; any plastic deformation that then occurred was fully irreversible. Again, the speed of load application or removal and the duration for which a given stress was applied did not influence the deformation that resulted.

For many situations, another type of deformation is noted and a one to one correspondence between stress and strain no longer prevails. Instead, the behavior is *time-dependent* and the resulting deformation is influenced by so-called viscous effects. Somewhat like plastic deformation, purely viscous effects may be permanent or irreversible but, unlike plastic deformation, are dependent upon time. Recall that true plastic deformation, as it has been defined, demands a minimum stress level to be exceeded; if stresses lower than that minimum are applied indefinitely, no plastic behavior occurs. With viscous deformation *any* applied stress will cause permanent deformation in time. This is the key difference between these two types of behavior.

To develop an acceptable constitutive relationship for such time-dependent solids, certain observed results must be satisfied. These include:

1. *Creep.* This is defined as an increase in strain with time as the stress or load is kept constant.* A full range of creep behavior would show that strain increases with time at a decreasing rate followed by a constant rate and, finally, an increasing rate.

2. *Recovery†* As the applied stress is reduced, either partially or completely, the strain diminishes with time. Terms such as inelastic, anelastic, and elastic aftereffect have also been used to describe this general behavior.

3. *Relaxation.* When a strain is induced and kept constant, the stress decreases with time.

4. *Rate effects.* The stress-strain curve rises more steeply as the applied strain rate or stress rate is increased.

These various effects are depicted in Fig. 6-1, and the major thrust of this chapter is to use models that satisfy these observed results at least in a qualitative manner. As with earlier models, a reasonable physical feel for the mechanics can be developed by using basic components to provide constitutive relations; they are intended to describe *linear viscoelastic behavior.* A more sophisticated approach, which is said to better describe the history of behavior, is discussed in detail by Williams,‡ yet no fully satisfactory approach has found universal acceptance at this time.

Polymers are often introduced as a class of solids that exhibit viscous behavior, probably because they display a time sensitivity under conditions of temperature and loading rate that cause negligible time sensitivity where

---

*In practice, such tests are usually conducted under constant load. To maintain a constant stress requires a coupling of load reduction with decreasing area.

†This has nothing to do with recovery of cold worked metals discussed in Chap. 7.

‡J. G. Williams, *Stress Analysis of Polymers* (New York: John Wiley and Sons, Inc., 1973) pp. 81–90.

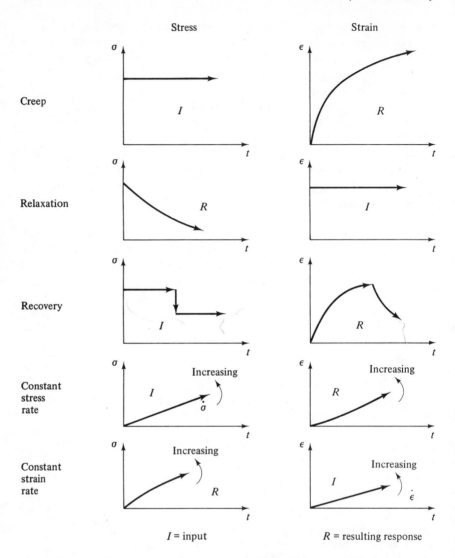

Stress   Strain

Creep

Relaxation

Recovery

Constant
stress
rate

Constant
strain
rate

I = input R = resulting response

Figure 6-1 Various effects of time-dependent behavior of stress and strain.

metals are involved. Such a sharp distinction could be misleading since metals will certainly display time-dependent behavior *if the testing or service conditions are of the proper magnitude.* It would be more correct to state that under equivalent conditions, polymers display a greater sensitivity to viscous effects than do metals. It is really a matter of degree and is due to the inherent structural differences of these two classes of solids. For our purposes, only the general macroscopic behavior involving viscous deformation is considered.

6-2 SIMPLE MODEL DEPICTING TIME DEPENDENT OR VISCOUS BEHAVIOR

To provide a reasonably complete overview, it would be most appropriate to discuss visco–elastic–plastic deformation since each of these distinct categories could be involved. Most sources limit such discussions to viscoelastic effects only; this is acceptable if the goal is to describe the various effects shown in Fig. 6-1. It is important to realize, however, that any permanent deformation arises solely because of viscous effects.*

Springs and friction elements were used earlier to model elastic and plastic effects. The component that describes purely viscous behavior is a simple dashpot as shown in Fig. 6-2. Note the following:

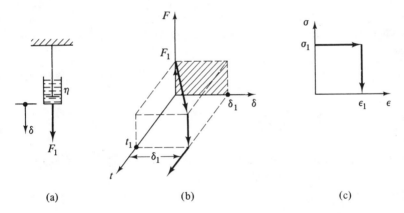

(a)   (b)   (c)

Figure 6-2 Time-dependent behavior of a linear, viscous solid showing the model (a), the force- displacement-time plot (b), and the stress-strain plot (c).

1. The dashpot contains a perfectly viscous fluid and the rate of extension or strain rate is directly proportional to the load or stress. Thus:

$$F = \eta \dot{\delta} \qquad (6\text{-}1)$$

where η is viewed as a material property.

2. With a sudden application of load, F_1, the dashpot acts as a rigid body and no immediate extension takes place at zero time. As the force is maintained, a linear extension results such that:

$$\delta_1 = \frac{F_1 t_1}{\eta} \qquad \text{after time } t_1 \qquad (6\text{-}2)$$

*Viscoelasticity is sometimes defined as deformation that fully recovers with time. This is not the concept followed in this book.

127

3. Load removal finds a permanent extension, δ_1, remaining so all of the input energy (shaded area) is dissipated during deformation, none being recovered Observe the similarity with plastic deformation shown in Fig. 4-1, but note that these types of irreversible deformation are due to different causes.

6-3 COMPOUND MODELS

By using the dashpot and other components in various combinations, a better portrayal of real behavior results. From here on, a change in symbolization will be used. Although the F-δ-t approach is more basic, it is the σ-ϵ-t plots that are of greater concern and one can interchange force with stress and extension with strain without any loss of generality. The behavior of an individual spring is given by $\sigma = E\epsilon$ but the elastic modulus of a compound model is not *necessarily* equal to the modulus of a single spring. Equations (6-1) and (6-2) may now be written as:

$$\sigma = \eta\dot{\epsilon} \quad \text{and} \quad \epsilon_1 = \frac{\sigma_1 t_1}{\eta} \tag{6-3}$$

6-4 MAXWELL MODEL

This model is formed by a spring and dashpot in series as shown in Fig. 6-3. The following considerations are pertinent:

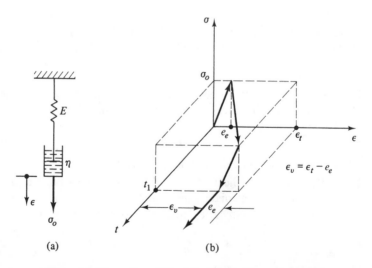

Figure 6-3 Maxwell model and the stress-strain–time behavior.

1. As σ_o is applied, an immediate elastic strain e_e results; the energy stored in the spring is recoverable.
2. Both Eqs. (6-3) and (3-1) are linear so the behavior of the dashpot and spring may be added directly using the principle of superposition.
3. After time t_1 has elapsed, a viscous strain, ϵ_v, is added, the total strain being composed of the two components. Note that the stress is common to both elements.
4. Complete removal of σ_o leads to elastic recovery of e_e but finds a permanent strain, ϵ_v, remaining.

To develop a constitutive relationship, the total strain under stress is ϵ_t where for time t_1,

$$\epsilon_t = e_e + \epsilon_v = \frac{\sigma_o}{E} + \frac{\sigma_o t_1}{\eta} \qquad (6\text{-}4)$$

In terms of rates, $\dot{\epsilon}_t = \dot{e}_e + \dot{\epsilon}_v$ and using Eqs. (3-1) and (6-3) for a uniaxial stress, there results

$$\frac{d\epsilon}{dt} = \frac{1}{E}\frac{d\sigma}{dt} + \frac{\sigma}{\eta} \qquad (6\text{-}5)$$

which assumes both E and η are *rate-independent*. To determine the appropriateness of this model, each of the basic requirements in Fig. 6-1 will be examined.

CREEP

With a constant value of σ_o, inspection of Eq. (6-4) shows there is an instantaneous strain, e_e, followed by an ever-increasing strain that varies linearly with time. This can also be deduced by viewing the ϵ-t plane in Fig. 6-3. A widely used term in this regard is the *creep compliance*, which is a function of time. It may be represented as $J(t)$, where

$$J(t) = \frac{\epsilon(t)}{\sigma_o} = \frac{1}{E}\left(1 + \frac{Et}{\eta}\right) \qquad (6\text{-}6)$$

This indicates that the strain at any time is proportional to the constant stress and is governed primarily by the term η/E. This ratio is called the *relaxation time* or *time constant* and is defined as $\tau = \eta/E$. The predicted creep behavior of a Maxwell model is not wholly realistic since the usual decreasing rate of strain with time, as observed with real solids, is not in evidence.

RECOVERY

Reference to Fig. 6-3 or Eq. (6-4) indicates that after some time t_1, full removal of the stress would cause an immediate recovery of the elastic component of strain. Regardless of the time that then elapses, the viscous component of strain is irreversible. This overall behavior is not truly descriptive of solids that display recovery and there is no need to carry this further.

RELAXATION

Here, a sudden strain,* say e_o, is induced and kept constant. Its magnitude is wholly dependent upon the elastic component since the dashpot acts as a rigid body under the initial application of stress. With the passage of time, the viscous component undergoes strain. This *must* be accompanied by a contraction of the elastic component if the overall strain is to remain fixed, so the stress decreases. Figure 6-4 shows a plot of this behavior and since at any time ϵ_o is constant, $d\epsilon/dt$ equals zero; Eq. (6-5) may be solved accordingly.

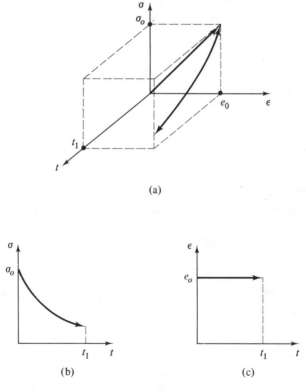

(a)

(b)   (c)

Figure 6-4 Stress relaxation of a Maxwell model.

*It is simpler to induce a sudden strain to show a graphical plot of relaxation; this is not essential.

Note that the initial *stress* σ_o must equal Ee_o at time $= 0$. Using this initial value,

$$\frac{d\sigma}{\sigma} = -\frac{E}{\eta}\,dt$$

so that

$$\ln \sigma = -\frac{Et}{\eta} + C_1 \qquad \text{where } C_1 = \ln \sigma_o \tag{6-7}$$

Thus,

$$\sigma = \sigma_o \exp\left(-\frac{Et}{\eta}\right) = \sigma_o \exp\left(-\frac{t}{\tau}\right)$$

or

$$\sigma = e_o E \exp\left(-\frac{t}{\tau}\right) \tag{6-8}$$

Here the relaxation modulus, which is the ratio of stress at a given time to the constant strain, is denoted as $E_r(t)$ where

$$E_r(t) = \frac{\sigma(t)}{e_o} = E \exp\left(-\frac{t}{\tau}\right) \tag{6-9}$$

This behavior provides a reasonable qualitative display of relaxation.

RATE EFFECTS*

First consider a constant rate of *stress* application such that in Eq. (6-5), $d\sigma/dt = A$. Then, $\sigma = At$ since the initial values are $\sigma = 0$ when $t = 0$.

$$\frac{d\epsilon}{dt} = \frac{A}{E} + \frac{At}{\eta} \tag{6-10}$$

The solution of Eq. (6-10) is:

$$\epsilon = \frac{At}{E} + \frac{At^2}{2\eta} \qquad \text{since } \epsilon = 0 \text{ at } t = 0 \tag{6-11}$$

An increase in A causes a nonlinear increase in strain, thereby raising the σ-ϵ curve with higher stress rates. Figure 6-5 illustrates this result.

For constant *strain* rates of application, $d\epsilon/dt = D$ and Eq. (6-5) becomes:

$$D = \frac{1}{E}\frac{d\sigma}{dt} + \frac{\sigma}{\eta} \tag{6-12}$$

Rearranging and using the integrating factor $\exp(Et/\eta)$ gives:

$$\sigma = \eta D\left[1 - \exp\left(-\frac{t}{\tau}\right)\right] \tag{6-13}$$

*Neither rate effects nor relaxation can be interpreted from Fig. 6-3; it is necessary to use the rate equations for a solution.

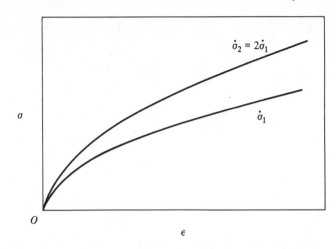

Figure 6-5 Influence on the stress-strain behavior as a function of stress rate using a Maxwell model.

since $\sigma = 0$ at $t = 0$ and $\tau = \eta/E$. This also raises the σ-ϵ curve with higher rates, but the stress tends to saturate with strain. Figure 6-6 shows the behavior.

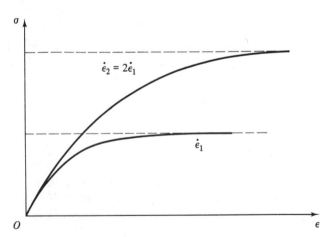

Figure 6-6 Influence on the stress-strain behavior as a function of strain rate using a Maxwell model.

6-5 VOIGT OR KELVIN MODEL

This model is formed by a spring and dashpot in a parallel configuration as shown in Fig. 6-7. Note that:

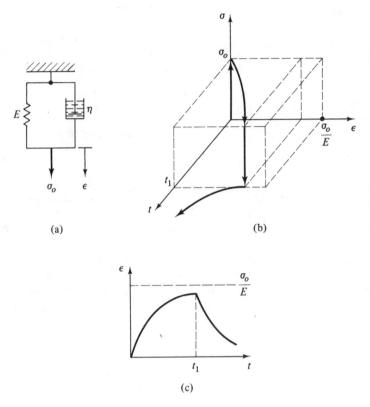

Figure 6-7 Kelvin model (a) showing the σ-ϵ-t behavior (b) and creep and recovery (c).

1. As a stress σ_o is suddenly applied, the dashpot prevents an instantaneous extension of the elastic component and each component supports a portion of σ_o (or the applied force).

2. With time, the viscous behavior causes an ever-increasing nonlinear strain and, with this model, the total strain, the strain sensed by the elastic component, and the viscous strain are all equal.

3. If the stress is removed, the elastic strain energy will begin to force movement of the viscous component and the recovery shows a nonlinear behavior with time.

A constitutive relationship for the Kelvin model is deduced from the stress relationships such that,

$$\sigma_o = \sigma_e + \sigma_v = Ee + \eta\dot{e}_v \qquad (6\text{-}14)$$

or

$$\frac{d\epsilon}{dt} = \frac{\sigma}{\eta} - \frac{E}{\eta}\epsilon \qquad (6\text{-}15)$$

since $\epsilon_t = e_e = \epsilon_v$, and E and η are again independent of time. Now consider the various behaviors predicted by this model, using ϵ for strain throughout.

CREEP

For an applied stress σ_o, Eq. (6-15) can be rearranged to give:

$$d\epsilon + \frac{E}{\eta}\epsilon\, dt = \frac{\sigma_o}{\eta}\, dt \qquad (6\text{-}16)$$

With E/η being $1/\tau$, ϵ being zero when $t = 0$, and using the integrating factor $\exp(t/\tau)$, the solution of Eq. (6-16) is:

$$\epsilon = \frac{\sigma_o}{E}\left[1 - \exp\left(-\frac{t}{\tau}\right)\right] \qquad (6\text{-}17)$$

As compared with Eq. (6-6), the predicted creep behavior of the Kelvin model is more realistic, since the strain approaches σ_o/E (i.e., the elastic strain if no dashpot were present) as time approaches infinity and does so at a decreasing rate. This is observed with many solids during the early stage of creep. Figure 6-7 shows this behavior.

The creep compliance function for this model is:

$$J(t) = \frac{\epsilon(t)}{\sigma_o} = \frac{1}{E}\left[1 - \exp\left(-\frac{t}{\tau}\right)\right] \qquad (6\text{-}18)$$

RECOVERY

Figure 6-7 shows that after time t_1 has elapsed, removal of σ_o would cause ϵ to approach zero in an exponential manner with the passage of time. This behavior is more qualitatively correct than that found with the Maxwell model and may be analyzed using the following derivation.[*]

Equation (6-17) shows that the strain caused by the applied stress is:

$$\epsilon = \frac{\sigma_o}{E}\left[1 - \exp\left(-\frac{t}{\tau}\right)\right] \qquad \text{for } 0 \leq t \leq t_1$$

If at time t_1, σ_o is suddenly reduced to zero, and expressing the continuation in time as $t' = t - t_1$ where t is the total time elapsed after the initial application of σ_o, then Eq. (6-16) becomes:

$$d\epsilon + \frac{\epsilon}{\tau}\, dt' = 0 \qquad \text{for } t \geq t_1 \qquad (6\text{-}19)$$

Using the initial value that $\epsilon = \epsilon_1$ when $t' = 0$, the solution of Eq. (6-19) is:

$$\epsilon = \epsilon_1 \exp\left(-\frac{t'}{\tau}\right) \qquad (6\text{-}20)$$

[*]David Ramsey, a former student, first suggested this derivation to the author.

Substituting $t - t_1$ for t' and Eq. (6-17) for ϵ_1 at time t_1, Eq. (6-20) reduces to:

$$\epsilon = \frac{\sigma_o}{E}\left[\exp\left(\frac{t_1}{\tau}\right) - 1\right]\exp\left(-\frac{t}{\tau}\right), \qquad t \geq t_1 \qquad (6\text{-}21)$$

Note that with Eq. (6-21), as $t \to t_1$, $\epsilon \to \epsilon_1$ while as $t \to \infty$ (with $t > t_1$), $\epsilon \to 0$.

RELAXATION

With the application of a constant strain, ϵ_o, then $d\epsilon/dt$ is zero and Eq. (6-15) shows:

$$\sigma_o = E\epsilon_o = \text{constant} \qquad (6\text{-}22)$$

In essence, the two components are locked together and there is no contraction of the elastic component with time as was evidenced with the Maxwell model. Thus, no relaxation is predicted by the Kelvin unit.

RATE EFFECTS

Consider a constant *strain* rate of application, $D = d\epsilon/dt$. Using Eq. (6-15) where $\epsilon = Dt$:

$$\frac{d\epsilon}{dt} = D = \frac{\sigma}{\eta} - \frac{E}{\eta}Dt \qquad (6\text{-}23)$$

since ϵ is zero at zero time. Thus,

$$\sigma = EDt + D\eta \qquad (6\text{-}24)$$

This implies an initial viscous response, which depends upon D, followed by a linear elastic increase of stress with strain. Figure 6-8 indicates the effect of two different constant strain rates on the resulting σ-ϵ behavior and the Maxwell model is more realistic in this regard.

For the application of a constant *stress* rate, $d\sigma/dt = A$, so $\sigma = tA$ since at zero time the stress is zero. Using this relation in Eq. (6-15),

$$\frac{d\epsilon}{dt} = \frac{A}{\eta}t - \frac{E}{\eta}\epsilon \qquad (6\text{-}25)$$

This can be rearranged to give:

$$d\epsilon + \frac{1}{\tau}\epsilon\,dt = \frac{A}{\eta}t\,dt$$

and using the integrating factor $\exp(t/\tau)$ plus the initial condition that ϵ is zero at zero time, the solution is:

$$\epsilon = \frac{A}{E}\left\{t - \tau\left[1 - \exp\left(-\frac{t}{\tau}\right)\right]\right\} \qquad (6\text{-}26)$$

Although an increase in stress rate does raise the σ-ϵ curve, the behavior is qualitatively correct only at relatively short times or small strains. Figure 6-9 displays the general result.

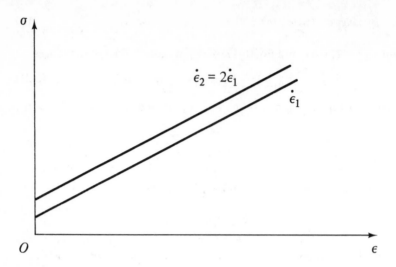

Figure 6-8 Effect of strain rate on the σ-ϵ behavior of a Kelvin model.

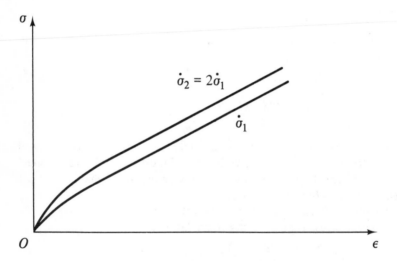

Figure 6-9 Effect of stress rate on the σ-ϵ behavior of a Kelvin model.

It has now been shown that a Maxwell model gives a reasonable prediction of relaxation behavior whereas a Kelvin model provides a better description for creep and recovery. Although both models lead to a raising of the σ-ϵ curve with an increase in stress or strain rate, strain rate effects are better described by the Maxwell model as are stress rate effects for large strains or long times. For small values of time, the Kelvin model displays reasonable behavior. By combining these models in various ways, even more realistic behavior is displayed. To demonstrate this, several compound models are now presented.

6-6 THREE-COMPONENT MODEL

Consider a spring in series with a Kelvin component as shown in Fig. 6-10(a) where the two springs have moduli of E_1 and E_2 respectively; the general loading and unloading behavior is indicated in Fig. 6-10(b). To develop the constitutive relationship, two relationships are used. First,

$$\epsilon = \epsilon_e + \epsilon_k \qquad (6\text{-}27)$$

where ϵ is the total strain and is composed of the sum of elastic and Kelvin strains. In addition,

$$\sigma = \sigma_1 + \sigma_2 \qquad (6\text{-}28)$$

where the applied stress is common to the individual spring and the Kelvin portion. This stress is distributed within the Kelvin portion as indicated in Fig. 6-10, the strain in this branch being the same for each component.

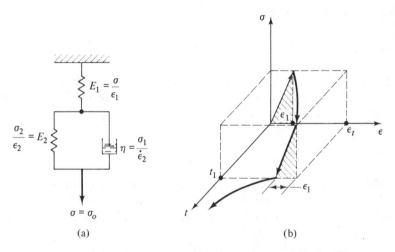

Figure 6-10 Three-component model (a) and the σ-ϵ-t behavior (b).

Letting the elastic strain be ϵ_1 and the Kelvin strain be ϵ_2, Eq. (6-27) in rate from becomes:

$$\frac{d\epsilon}{dt} = \frac{d\epsilon_1}{dt} + \frac{d\epsilon_2}{dt} = \frac{1}{E_1}\frac{d\sigma}{dt} + \frac{d\epsilon_2}{dt} \qquad (6\text{-}29)$$

Equation (6-28) can be written as

$$\sigma = \eta\frac{d\epsilon_2}{dt} + \epsilon_2 E_2 = \eta\frac{d\epsilon_2}{dt} + E_2(\epsilon - \epsilon_1)$$

or

$$\frac{d\epsilon_2}{dt} = \frac{1}{\eta}\left(\sigma - E_2\epsilon + \frac{\sigma E_2}{E_1}\right) \qquad (6\text{-}30)$$

Combining Eqs. (6-29) and (6-30) gives,

$$\frac{d\epsilon}{dt} + \frac{E_2}{\eta}\epsilon = \frac{1}{E_1}\frac{d\sigma}{dt} + \frac{\sigma}{\eta}\left(1 + \frac{E_2}{E_1}\right) \tag{6-31}$$

CREEP

With $d\sigma/dt = 0$, ϵ being zero at zero time, and following procedures such as those which produced Eq. (6-18) the creep compliance function is found to be:

$$J(t) = \frac{1}{E_1} + \frac{1}{E_2}\left[1 - \exp\left(-\frac{t}{\tau}\right)\right] \tag{6-32}$$

where $\tau = \eta/E_2$. Note that by using superposition as in Eq. (6-27), the creep compliance function follows directly since

$$\epsilon = \epsilon_1 + \epsilon_2 = \frac{\sigma_o}{E_1} + \frac{\sigma_o}{E_2}\left[1 - \exp\left(-\frac{t}{\tau}\right)\right]$$

which is equivalent to Eq. (6-32).

RECOVERY

Figure 6-10 indicates that removal of the stress σ_o causes an immediate recovery of the elastic strain; the subsequent behavior is identical to the Kelvin model as given by Eq. (6-21).

RELAXATION

With a constant applied strain, ϵ_o, $d\epsilon/dt = 0$, and the relaxation modulus can be found using a procedure similar to that which led to Eq. (6-9). Again, when $t = 0$, $\sigma = \sigma_o = \epsilon_o E_1$. The result can be expressed as

$$E_r(t) = E_1 - \frac{E_1^2}{E_1 + E_2}\left[1 - \exp\left(-\frac{t}{\tau}\right)\right] \tag{6-33}$$

where $\tau = \eta/(E_1 + E_2)$.

RATE EFFECTS

With a constant stress rate, $A = d\sigma/dt$, and stress being zero when $t = 0$, $\sigma = tA$ for any time. Following earlier procedures, the solution of Eq. (6-31) is:

$$\epsilon = \frac{At}{E_1} + \frac{A}{E_2}\left\{t - \tau\left[1 - \exp\left(-\frac{t}{\tau}\right)\right]\right\} \tag{6-34}$$

where $\tau = \eta/E_2$. Note that the first term on the right-hand side is the elastic strain as a function of time while the second term is identical to Eq. (6-26).

The derivation for σ as a function of time based upon a constant strain rate, $d\epsilon/dt = D$, is left as an exercise.

6-7 FOUR-COMPONENT MODEL

This model consists of a Maxwell unit in series with a Kelvin unit as shown in Fig. 6-11 along with the stress-strain-time plot. Note the following:

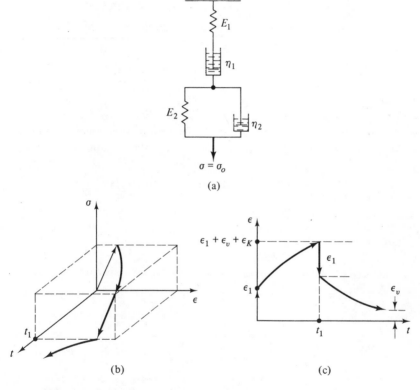

Figure 6-11 Four-component model (a), the σ-ϵ-t behavior (b), and creep and recovery (c).

where E_1 is the modulus of the single spring, E_2 is the modulus of the spring in the Kelvin portion and $\tau = \eta/(E_1 + E_2)$. This is Eq. (6-33).

6-10 The behavior of a viscoelastic solid might be described by the model shown.
(*a*) Derive the governing differential equation in rate form for this model.

1. As σ_o is suddenly applied, the spring having modulus E_1 provides an instantaneous elastic strain.

2. The time-dependent strain is composed of the viscous strain due to η_1 and the strain displayed by the Kelvin unit (E_2 and η_2).

3. Removal of σ_0 causes an instantaneous elastic recovery followed by a time-dependent recovery that, in the limit, approaches the irreversible strain due to η_1.

In essence, the Maxwell portion governs elastic effects, relaxation, and permanent viscous strain while the Kelvin portion controls recovery. Creep and rate effects are influenced by both; thus, the full spectrum of material response is better satisfied when compared with either model acting individually. With the use of superposition, the total strain is simply the sum of the individual components and the result is:

$$\epsilon = \frac{\sigma_o}{E_1} + \frac{\sigma_o t}{\eta_1} + \frac{\sigma_o}{E_2}\left[1 - \exp\left(-\frac{t}{\tau_2}\right)\right] \tag{6-35}$$

Figure 6-11(c) displays this combined effect.

As with the earlier developments, the responses of recovery, relaxation, and rate effects may be determined using the four component model; this forms a series of exercises for the reader. Many other combinations could be introduced in order to produce subtle refinements, and with the background of this chapter, the reader should be able to handle such modifications where that is of interest.

PROBLEMS

6-1 Sketch the deflection versus time plot of a Kelvin body subjected to the following:
   (a) $F = F_1$, $0 \leqq t \leqq t_1$.
   (b) $F = 0$, $t_1 \leqq t \leqq t_2$.
   (c) $F = -F_1$, $t_2 \leqq t \leqq t_3$.
   Show your δ-t plot in correspondence with the F-t plot, noting that F and δ are proportional respectively to σ and ϵ.

6-2 For the model (p. 141) shown, develop an expression for deflection, δ, as a function of E, η, and t. Note that the block, being a friction element, will not move until it senses a force greater than F_o. This model is called a Bingham body.

6-3 Based upon the four-element model shown in Fig. 6-11, explain why:
   (a) Elastic moduli measured at different strain rates have different values.
   (b) Elastic moduli measured at different temperatures have different values.
   (c) Constant applied stress causes continuing deformation.

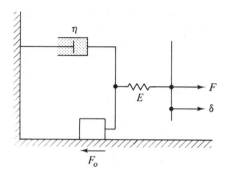

6-4 Comparing a Maxwell versus a Kelvin model:
    (*a*) Which better describes creep? Explain.
    (*b*) Which better describes relaxation? Explain.

6-5 Consider a Maxwell model that is subjected to two constant strain rates where $\dot{\epsilon}_2 = 2\dot{\epsilon}_1$. By computing enough points, show how the σ-ϵ (i.e., stress-strain curve) for $\dot{\epsilon}_2$ compares with $\dot{\epsilon}_1$. These should be plotted to scale! *Hint*: Consider η and τ are both unity for purposes of simplicity.

6-6 Repeat Prob. 6-5 using a Kelvin model.

6-7 For the three-element model shown in Fig. 6-10, it is to be assumed that this gives a reasonable representation of the behavior of a certain solid. When a load of 1000 lbf is suddenly applied, the instantaneous strain is 0.020. The initial unloaded length and diameter are 5 in. and 1 in. respectively.

    After one minute the strain has increased to a value of 0.049 while after 10 minutes it has increased to a total value of 0.095. From the instant the constant load was applied, how much time would elapse before the specimen was 6 in. long? *Note*: Do not use a graphical approach although you may *check* your answer by doing so.

6-8 Derive the governing differential equation in rate form for a four-element model (Maxwell in series with Kelvin) then show that a creep analysis (where $d\sigma/dt = 0$) leads to the creep compliance function:

$$J(t) = \frac{\epsilon(t)}{\sigma} = \frac{1}{E_1} + \frac{t}{\eta_1} + \frac{1}{E_2}\left[1 - \exp\left(-\frac{tE_2}{\eta_2}\right)\right]$$

where E_1 and η_1 pertain to the Maxwell components while E_2 and η_2 pertain to the Kelvin components. *Hint*: Procedures similar to those that led to Eq. (6-32) should be used.

6-9 For the three-component model, show that the relaxation modulus may be expressed as:

$$E_r(t) = \frac{\sigma(t)}{\epsilon} = E_1 - \frac{E_1^2}{E_1 + E_2}\left[1 - \exp\left(-\frac{t}{\tau}\right)\right]$$

(b) Determine the relaxation of stress σ as a function of the applied stress σ_o, time t, and the material constants η_1, η_2 and E.

(c) By using the concept of superposition, draw a neat free-hand sketch of strain as a function of time when such a model is stressed for some period and then is fully unloaded. Be sure to indicate what happens to ϵ as t approaches infinity.

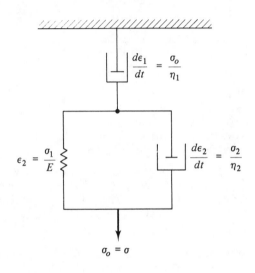

$$\left| \frac{d\epsilon_1}{dt} = \frac{\sigma_o}{\eta_1} \right.$$

$$\epsilon_2 = \frac{\sigma_1}{E}$$

$$\left| \frac{d\epsilon_2}{dt} = \frac{\sigma_2}{\eta_2} \right.$$

$$\sigma_o = \sigma$$

7

ELEMENTS OF
DISLOCATION THEORY

7-1 INTRODUCTION

Both elastic and plastic deformation have been discussed on a macroscopic scale in earlier chapters. At the atomic level, which shall be referred to as the microscopic approach, it has been found that plastic behavior cannot be described as a simple extension of elastic deformation. If the usual hard-ball model is used to depict atomic structure, and if the solid is crystalline in nature, *as with metals*, then the model describing the force reactions between two atoms serves as a reasonable starting point. With such a model, the atoms are positioned at some equilibrium spacing that results from the balance of attractive and repulsive forces which typify the metallic type of bonding. The application of a small external force tends to move the atoms from this equilibrium position by extremely minute amounts until a new condition of force equilibrium occurs. Removal of the external force then sees a restoration of the original atomic alignment due to the balancing of the attractive and repulsive forces that comprise atomic bonding. In such a context, elastic responses may be understood to be a direct function of these atomic forces and quite accurate calculations of such responses have been made.

Concerning plastic behavior on a microscopic scale, one of the primary causes is due to the mechanism called *slip*.* In the most general case, slip occurs on planes of greatest atomic density in the direction of greatest linear density. As a consequence, slip is quite anisotropic in behavior; this is certainly true in cases where the deformation of single crystals is involved. Although slip leads to a relative displacement of atoms, the basic crystalline structure is unchanged.

For many years, it was assumed that slip occurred by having full planes of atoms move simultaneously with respect to adjacent planes. Various analytical approaches to this problem all led to the conclusion that the shear strength of real metals was far lower than it should be if the manner in which slip occurred was by full plane movement. In 1934, Taylor [4], Orowan [5], and Polanyi [6], apparently working independently, all postulated the concept that a certain type of defect in a crystalline structure might account for the great discrepancy between theoretical calculations and observed behavior. This defect has been termed a *dislocation* and led to what is commonly referred to as dislocation theory. The theory seemed to be quite rational since it answered a number of puzzling questions, but it was not until the early 1950's, especially with the aid of the transmission electron microscope (TEM), that any extensive support of the existence of dislocations was forthcoming.† On this basis it is essential to consider the fine details of crystalline structure in regard to plasticity. Such is not the case in elastic analyses, and for that reason, plasticity cannot be considered to be a simple extension of elasticity.

It may be concluded that a satisfactory theory of microplasticity must include at least the following requirements:

1. The material remains crystalline after slip has occurred.
2. Slip occurs on planes in some type of consecutive behavior, so full planes of atoms do not slip simultaneously.
3. Slip starts at some irregularity or defect in the crystalline structure.

7-2 MAXIMUM THEORETICAL SHEAR STRESS

To indicate why it was necessary to postulate a concept such as dislocation theory, a calculation of the shear stress that would be necessary to cause slip by the simultaneous movement of full atomic planes is instructive.

*Twinning is not discussed in this text. Other references [1–3] may be consulted.
†The etch-pit technique was used earlier.

In Fig. 7-1 a portion of two adjacent planes of atoms is depicted in equilibrium. There, each atom touches its neighbors which implies that an atom such as A rests in the valley formed by B and C. Slip is assumed to occur by having all atoms on plane D move simultaneously with respect to plane E due to some applied shear stress, τ.

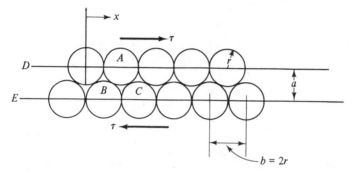

Figure 7-1 Adjacent rows of atoms using an idealized hard-ball model.

As τ is applied and, for simplicity, holding the atoms on plane E fixed, the atoms on D will undergo a displacement in the x direction. If τ is small and cannot fully overcome the bonding forces, the displacement is considered as elastic in nature. Removal of τ would see a return of all atoms to their original positions. If the shear stress causes a displacement of $\frac{1}{2}b$, every atom on D would be directly above and aligned with an atom on E and, because of this symmetry, the removal of τ could find the atoms moving as easily to the right as to the left in order to return to an equilibrium condition. Thus, when x equals $\frac{1}{2}b$, τ can be taken as zero whereas when x equals b, the initial atomic configuration has returned but all atoms on D have *slipped* a unit distance b or twice the atomic radius r. In order to cause this slip, some value τ_{max} must be applied between the displacement $0 < x < \frac{1}{2}b$. The variation of τ with displacement x is most often assumed to be a sine function (this is *not* the only form used in this context); in essence,

$$\tau = \tau_{max} \sin\left(\frac{2\pi x}{b}\right) \tag{7-1}$$

and this is described by Fig. 7-2.

If x equals zero or $\frac{1}{2}b$, τ is zero and if x equals $\frac{1}{4}b$, τ reaches its maximum value. Considering the case where x is small (i.e., $x \ll b$), the slope of the plot can be taken as τ/x. In addition, using Eq. (7-1) gives:

$$\frac{d\tau}{dx} = \tau_{max}\frac{2\pi}{b}\left(\cos\frac{2\pi x}{b}\right) \approx \frac{2\pi\tau_{max}}{b} \tag{7-2}$$

for small x since $\cos\left(\frac{2\pi x}{b}\right) \approx 1$.

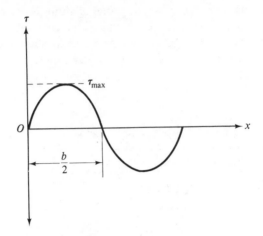

Figure 7-2 One possible relationship between applied shear stress and displacement.

Equating Eq. (7-2) with the expression for the slope at small x:

$$\frac{2\pi\tau_{max}}{b} = \frac{\tau}{x} \qquad (7\text{-}3)$$

Now $\tau = G\gamma = G(x/a)$ where G is the shear modulus and γ is the shear strain. Since a and b are of the same magnitude:

$$\tau \approx G\left(\frac{x}{b}\right) \qquad (7\text{-}4)$$

Combining Eqs. (7-3) and (7-4) results in:

$$\tau_{max} \approx \frac{G}{2\pi} \qquad (7\text{-}5)$$

For most metals, the value of τ_{max} is on the order of 400 to 1000 times greater than that actually observed. Even using more refined analyses [7], the discrepancy between theoretical predictions and real behavior is too great to ignore. From this it must be concluded that real solids do not possess perfect crystalline structure and that slip cannot take place by the simultaneous shearing of full atomic planes. The concepts behind dislocation theory provide certain rational explanations in this regard.

Example 7-1.

The measured shear stress to cause slip in pure aluminum is about 1000 psi (6.859 MPa). What is the ratio of this actual value to the theoretical maximum shear stress to cause slip?

Solution.

The elastic modulus of aluminum is on the order of 10^7 psi (68.95 GPa) and Poisson's ratio is about 0.33. Therefore, the shear modulus (assuming isotropy) is, from Eq. (3-7):

$$G = \frac{E}{2(1+v)} = \frac{10^7}{2.66} = 3.76 \times 10^6 \text{ psi} = 25.9 \text{ GPa}$$

From Eq. (7-5), $\tau_{\max} = 25.9/(2\pi) = 4.13$ GPa. Thus, 4.13 GPa/6.895 MPa = 600, so the value of $\tau_{\max}$ is about 600 times the experimentally observed value.

7-3 MODELS OF PURE EDGE AND PURE SCREW DISLOCATIONS

Various models have been used to portray what are called *pure edge, pure screw,* and *mixed* dislocations. The models shown in Fig. 7-3 may prove to be the most instructive, especially when used in connection with subsequent discussions. It must be understood that actual dislocations are not formed in this manner, but the physical results caused by the existence of dislocations may be illustrated with the use of these models.

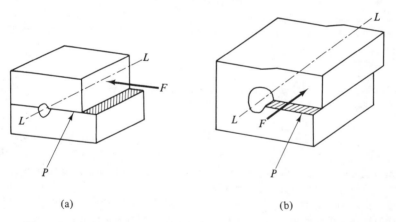

(a)   (b)

Figure 7-3 Conceptual model showing formation of an edge dislocation (a) and a screw dislocation (b).

Figure 7-3(a) portrays a block in which a hole is drilled. The top of the block is then displaced due to the force F and the interface at P is glued. If F is removed, the top half cannot restore itself to its position prior to the force application—the atomic arrangement, especially in the vicinity of the line LL, has been distorted from its equilibrium configuration. The end result has led to an *edge dislocation* where the region above the dislocation line is in residual compression while that below feels residual tension. In a similar manner, Fig. 7-3(b) shows a *screw dislocation*; there the top half of the block has been displaced by a force acting parallel to *LL*. To produce either result required an input of work or energy to displace the atoms from equilibrium and produce a dislocation. Thus, the presence of a dislocation causes the strain energy of a crystal to be higher than that which would exist if the dislocation were not present.

Other forms of dislocation models are illustrated in Figs. 7-4 and 7-5. Note that the edge dislocation produces an extra *half-plane* of atoms and the connection of all atoms at the bottom of the half plane may be envisioned as forming a line such as *LL* in Fig. 7-4(a). For this reason, dislocations are called *line defects* and in later analyses, it is helpful to consider them in this context. With a screw dislocation, there is no extra half-plane of atoms; instead, if one traversed through the crystal structure by circling the dislocation line [*LL* in Fig. 7-4(b)], a helical or screwlike path would be followed where the pitch for each 360 degrees rotation would be constant. It is somewhat surprising that although screw dislocations are a bit more difficult to visualize—and making physical models for display brings this out—mathematical descriptions of screw dislocation behavior are much simpler than the comparable descriptions for the edge type. In reality, most dislocations are of a mixed nature, that is, neither pure screw nor pure edge for the full dislocation line. Figure 7-6 illustrates this, where the part at *C* is pure edge, at *A* is pure screw, and the remainder is mixed.

7-4 DISLOCATION MOTION AND THE BURGERS VECTOR

Reference to the edge model in Fig. 7-4 provides an inkling as to why the theoretical τ_{max} is not attained. Rather than full planes of atoms slipping simultaneously, which would require the instantaneous severing of many atomic bonds, the dislocation is seen to move under an applied stress as illustrated schematically in Fig. 7-7. In essence, relatively few bonds need be broken or overcome at any instant thereby demanding a much lower shear stress to cause the process called slip. Such movement, which is a result of the *easy glide* of dislocations, requires that $\tau_{applied} \ll \tau_{max}$ for slip to take place. Note that if a single dislocation moved through the full crystal, the atomic rearrangement would lead to a perfect structure since the line defect has been removed. Of course, real crystalline solids may possess millions of dislocations and their interactions can lead to phenomena far more complex than that depicted in Fig. 7-7.

In principle, a line could be drawn around the boundary that separates the slipped and unslipped regions of a crystal at any instant; this is the dis-

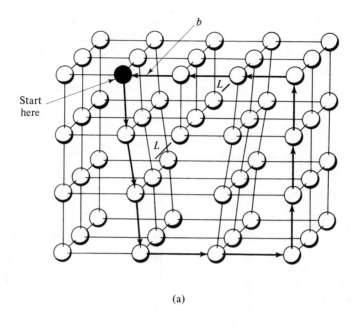

(a)

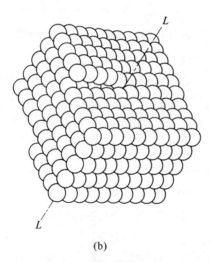

(b)

Figure 7-4 Hard-ball models of an edge dislocation (a) and a screw dislocation (b).

149

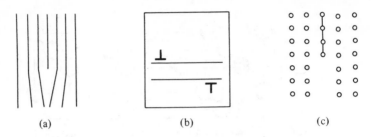

(a)   (b)   (c)

Figure 7-5 Various representations of an edge dislocation.

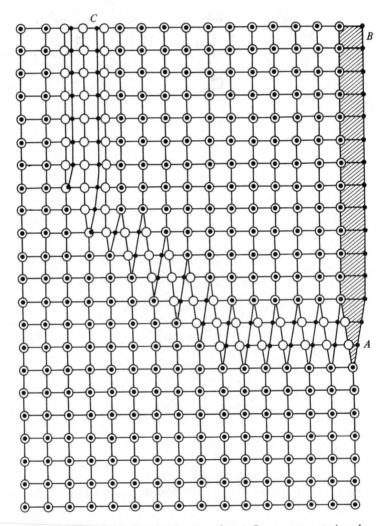

Figure 7-6 Dislocation line that is pure edge at C, pure screw at A, and mixed in the remainder.

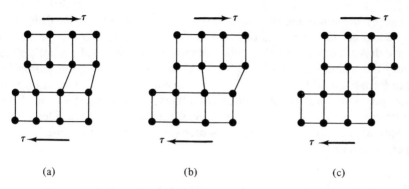

(a)   (b)   (c)

Figure 7-7 Consecutive slip in a crystal containing an edge dislocation.

location *line*. Figure 7-8 portrays such a case where in region *A* slip has occurred while in *B* it has not. EFGH is the dislocation line and may be defined by a vector which specifies the direction and so-called strength of the dislocation. This is the most characteristic feature of the dislocation and is called the *Burgers vector* after J. M. Burgers [8] who also proposed the concept of a screw dislocation. In Fig. 7-4 the manner of defining the Burgers vector is indicated for the pure edge dislocation. By tracing a circuit that surrounds the dislocation line, the distance needed to complete the circuit defines this vector. In the case of a perfect structure, a full circuit finds a return to the same starting point; no closing up is required since no dislocation exists in that portion of the crystal.* A similar approach is used for a screw dislocation. Throughout this chapter, the magnitude of the Burgers vector, *b*, equals twice the atomic radius of the solid under consideration. With reference to Fig. 7-1, this implies that an incremental amount of slip

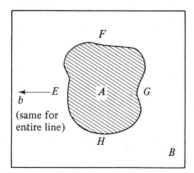

Figure 7-8 Dislocation line separating slipped (shaded) and unslipped areas on a slip plane.

*Whether the circuit is taken clockwise or the reverse depends upon the direction in which the sense of the dislocation is chosen as positive. See Figs. 7-9 and 7-12.

involves one unit of atomic displacement. For such situations, the dislocation is sometimes referred to as one of unit strength. (Note that it is possible to have dislocations of *less* than unit strength; these are called partial dislocations but will not be pursued further.) As seen in Fig. 7-8, a dislocation line can form a closed loop within a grain or crystal, but such a line cannot end abruptly within a grain. Any dislocation line has one Burgers vector of constant magnitude and direction; in this context the vector is considered to be invariant. Finally, the direction of *movement* of a dislocation line is always perpendicular to the line itself.

7-5 MATHEMATICAL DERIVATION OF STRESSES AND STRAINS CAUSED BY DISLOCATIONS

Since the presence of dislocations causes the lattice structure to be disrupted from equilibrium spacing, the regions of the crystal, especially near the dislocation, experience strains and stresses that would not otherwise exist. As the displacements involved are small, certain equations from elasticity theory may be used and, if acceptable displacement functions are found, the strains and stresses caused by the presence of a dislocation can then be determined. It should be realized at the outset that these expressions are held to be reasonable at all points *except* near the *core* of the dislocation (i.e., the dislocation line itself). In that region the equations would predict infinite strain or stress and are therefore invalid at or very near the core. Once the stress equations are developed, they can be used to determine the magnitude and sign of the forces that interact between dislocations. An explanation of certain observed macroscopic behavior is then possible.

DISPLACEMENTS AND STRAINS

With certain assumptions and constraints, the expressions for strains as a function of displacements were developed to give Eqs. (2-3) and (2-4). Normal strains were expressed as:

$$e_x = \frac{\partial u}{\partial x}, \qquad e_y = \frac{\partial v}{\partial y}, \qquad \text{and} \qquad e_z = \frac{\partial w}{\partial z} \qquad (7\text{-}6)$$

while shear strains were:

$$\gamma_{xy} = \frac{\partial v}{\partial x} + \frac{\partial u}{\partial y}, \qquad \gamma_{yz} = \frac{\partial w}{\partial y} + \frac{\partial v}{\partial z} \qquad \text{and} \qquad \gamma_{zx} = \frac{\partial u}{\partial z} + \frac{\partial w}{\partial x} \qquad (7\text{-}7)$$

GENERALIZED STRESS-STRAIN RELATIONS

The generalized form of Hooke's law for linear elastic deformation was discussed in Chap. 3 and the following were developed:

$$e_x = \frac{\sigma_x}{E} - \frac{v}{E}(\sigma_y + \sigma_z) \quad \text{and} \quad \gamma_{xy} = \frac{\tau_{xy}}{G} \tag{7-8}$$

Expressions for e_y, e_z, γ_{yz}, and γ_{zx} are indentical except for the proper interchange of subscripts. Recall that the shear modulus, elastic modulus, and Poisson's ratio were interrelated for an isotropic solid as follows:

$$G = \frac{E}{2(1 + v)} \tag{7-9}$$

For simplicity, it is useful to introduce the Lamé constant, λ, which is defined as:

$$\lambda = \frac{vE}{(1 + v)(1 - 2v)} \tag{7-10}$$

Although the use of λ shortens many expressions, it appears to have no distinct physical meaning. Note too that Eqs. (7-9) and (7-10) may be used to give the relationship $v = \lambda/(2\lambda + 2G)$.

In the developments that follow, it is helpful to revise Eq. (7-8) to give an explicit expression for the normal stress as a function of strain, the shear modulus, and the Lamé constant. The results are as follows:

$$\sigma_x = (\lambda + 2G)e_x + \lambda(e_y + e_z)$$
$$\sigma_y = (\lambda + 2G)e_y + \lambda(e_x + e_z) \tag{7-11}$$
$$\sigma_z = (\lambda + 2G)e_z + \lambda(e_x + e_y)$$

The derivation of Eq. (7-11) is not at all obvious so, for completeness, it is given in Appendix 7-A at the end of this chapter.

EQUILIBRIUM EQUATIONS

In Chap. 1 the stress equilibrium equations were derived from a force balance approach. One such equation was:

$$\frac{\partial \sigma_x}{\partial x} + \frac{\partial \tau_{yx}}{\partial y} + \frac{\partial \tau_{zx}}{\partial z} = 0 \tag{7-12}$$

Two other identical expressions would result with an appropriate interchange of subscripts.

Equations (7-6) through (7-12) provide the basis for further analysis. Once particular displacement functions are postulated, the resulting strains can be expressed through Eqs. (7-6) and (7-7). Stresses may then be determined with the use of Eq. (7-11) and checked to satisfy equilibrium with Eq. (7-12). To complete such an analysis, pertinent boundary conditions must also be satisfied.

A Special Case

As its usefulness will become apparent, it is convenient to express some of the previous equations for the case where there is no *variation* of displacement in one direction. Choosing the z direction as such, all derivatives with respect to z must vanish. A further development of great use is to rewrite the equilibrium equations in terms of displacements and elastic constants only. The following relations result:

$$(\lambda + 2G)\frac{\partial^2 u}{\partial x^2} + (\lambda + G)\frac{\partial^2 v}{\partial x\,\partial y} + G\frac{\partial^2 u}{\partial y^2} = 0$$

$$(\lambda + 2G)\frac{\partial^2 v}{\partial y^2} + (\lambda + G)\frac{\partial^2 u}{\partial x\,\partial y} + G\frac{\partial^2 v}{\partial x^2} = 0 \qquad (7\text{-}13)$$

$$G\left(\frac{\partial^2 w}{\partial x^2} + \frac{\partial^2 w}{\partial y^2}\right) = 0$$

To obtain Eq. (7-13), Appendix 7-B may be consulted as to the method involved. With Eq. (7-13) any postulated displacement functions can be tested quickly to determine if stress equilibrium is satisfied. Although these equations are correct for any situation where all derivatives with respect to z are zero, they will be used here in analyses connected with pure edge and pure screw dislocations only.†

Screw Dislocation

If it is accepted that the elasticity equations leading to Eq. (7-13) provide a reasonable description of the influence of a dislocation in a real solid, then the displacement at any point in the solid represents the displacement of atoms caused by the dislocation itself. There are various conventions used to describe a screw dislocation in relation to a chosen coordinate system and to its sense. A right-hand rule convention is used in this text. Figure 7-9 shows the path that traverses one full revolution around a screw dislocation whose line is zz. If the positive sense is chosen in the positive z direction, the result describes a left-handed screw dislocation whose Burgers vector is positive in the plus z direction and is *parallel* to line zz. Note that if a right-hand rule is used with the thumb pointing in the direction of positive sense, the fingers would then be turned in the direction of a left-handed thread; hence the dislocation is left-handed. If the pitch is in the opposite direction, a right-handed dislocation results. For the type of problems presented at the end of this chapter, it is not necessary to become overly concerned about such descriptions since it will always be stated whether the dislocations involved are of like or unlike sign (i.e., left- or right-handed). Figure 7-3(b) indicates

---

†The developments that follow relate to stationary dislocations.

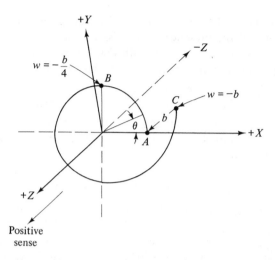

Figure 7-9 Displacement caused by a left-hand screw dislocation showing the dislocation line *ZZ* and the Burgers vector *b*.

that the only displacement involved here is in the *z* direction and this displacement, *w*, is constant everywhere along *zz* for any specific angular rotation. Starting at a point such as *A* in Fig. 7-9, the displacement for a full revolution or circuit to point *C* shows that $w = -b$. If only a 90° traverse were made to point *B*, $w = -b/4$. Thus a reasonable description of displacement *w* could be:

$$w = -b\left(\frac{\theta}{2\pi}\right) \tag{7-14}$$

or

$$w = -\frac{b}{2\pi}\tan^{-1}\frac{y}{x} \tag{7-15}$$

Now $u = v = 0$ since there is no displacement in the *x* and *y* directions, so Eq. (7-13) can be used to determine if these three displacement functions satisfy stress equilibrium. It is obvious that the first two relationships in Eq. (7-13) are satisfied since *u* and *v* are both zero. As to the third equation in that group, equilibrium demands that

$$\frac{\partial^2 w}{\partial x^2} = -\frac{\partial^2 w}{\partial y^2} \tag{7-16}$$

If the differentiation is carried out using Eq. (7-15), the result is:

$$\frac{\partial^2 w}{\partial x^2} = \frac{-2yxb}{2\pi(x^2 + y^2)^2} \quad \text{and} \quad \frac{\partial^2 w}{\partial y^2} = \frac{+2yxb}{2\pi(x^2 + y^2)^2} \tag{7-17}$$

so that the assumed displacement functions satisfy equilibrium and are acceptable on that basis. From Eqs. (7-6) and (7-7) all normal strains are found to be zero as is the shear strain γ_{xy}. Thus only the shear strains γ_{yz} and γ_{zx} exist because of the presence of the dislocation. This should not be too surprising since as shown in context with Fig. 7-3(b), a shear effect produced such a dislocation. As to the shear strains themselves, they are expressed by:

$$\gamma_{yz} = -\frac{b}{2\pi}\left(\frac{x}{x^2 + y^2}\right) \quad \text{and} \quad \gamma_{zx} = \frac{b}{2\pi}\left(\frac{y}{x^2 + y^2}\right) \qquad (7\text{-}18)$$

Using Eq. (7-8) for shear stresses and Eq. (7-11) for normal stresses, all normal stresses are found to be zero as is the shear stress τ_{xy}. The only non-zero shear stresses are:

$$\tau_{yz} = -\frac{Gb}{2\pi}\left(\frac{x}{x^2 + y^2}\right) \quad \text{and} \quad \tau_{zx} = \frac{Gb}{2\pi}\left(\frac{y}{x^2 + y^2}\right) \qquad (7\text{-}19)$$

In a physical sense, Eq. (7-18) describes the elastic strains induced in the crystal because of the presence of the dislocation while Eq. (7-19) provides the accompanying stresses. Because there is radial symmetry identified with the expression for the displacement w, it is often more convenient to express the shear strains and stresses in cylindrical coordinates as follows:

$$\gamma_{z\theta} = -\frac{b}{2\pi r} \quad \text{and} \quad \tau_{z\theta} = -\frac{Gb}{2\pi r} \qquad (7\text{-}20)$$

Figure 7-10 displays stress states at certain locations in terms of the cartesian system while Fig. 7-11 illustrates the results for cylindrical coordinates.

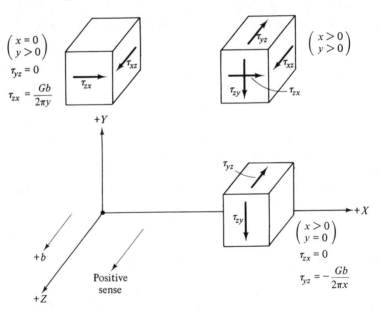

Figure 7-10 Stress states at various locations caused by a screw dislocation (Cartesian coordinates).

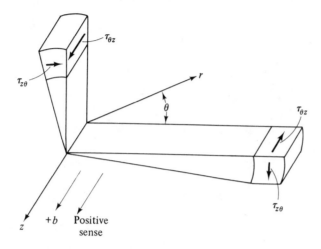

Figure 7-11 Same as Fig. 7-10 using cylindrical coordinates.

A very important point should be noted in regarding Eq. (7-20). Since neither strain nor stress is a function of θ, a screw dislocation is not identified with any particular slip plane. For this reason, the phenomenon called cross-slip can occur where screw dislocations are concerned. This is discussed in Sec. 7-11. To summarize the developments in this section, a chosen set of displacements was found to satisfy equilibrium and the most useful form of the only nonzero strains and stresses caused by the existence of the dislocation is given by Eq. (7-20). Although boundary conditions have been neither specified nor checked, it will suffice to say that they would not lead to any alteration of the final results presented above.*

EDGE DISLOCATION

Although simpler to portray schematically, the development of the mathematical description of the strains and stresses caused by an edge dislocation is more involved than that given for a screw dislocation. Figure 7-12 shows a sketch of an edge dislocation where the extra half-plane of atoms lies in a y-z plane in the negative y direction. Choosing the sense of the dislocation in the positive z direction results in a Burgers vector that is perpendicular to the dislocation line zz and is positive in the plus x direction. Using a right-hand rule, with the thumb in the direction of positive sense, means that the Burgers circuit is taken counterclockwise.

Figure 7-13 indicates how the atoms might be shifted to cause the disloca-

*See reference [9] for further details.

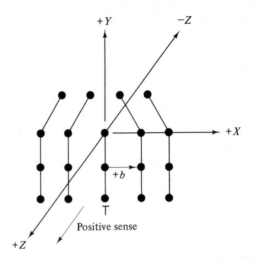

Figure 7-12 Representation of an edge dislocation showing the dislocation line zz and the Burgers vector b.

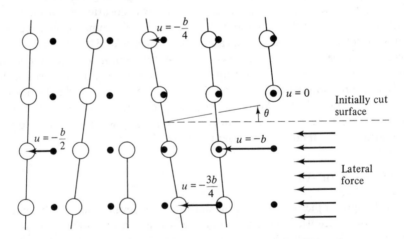

Figure 7-13 Possible displacements caused by an edge dislocation.

tion shown in Fig. 7-12; again, consider the application of a force acting against the lower part of the crystal in a manner analogous to Fig. 7-3(a). In Fig. 7-13 the small solid circles indicate the atomic arrangement before the dislocation exists whereas the large open circles are the atomic positions after dislocation formation. Note too that only the atoms in the vicinity of the dislocation are considered. As a first guess of acceptable displacement functions, it would appear that w and v, displacements in the z and y directions, are zero and that u might have a form identical to Eq. (7-14). If these functions are used, it is found that Eq. (7-13) is not fully satisfied so other forms of the displacement functions must be investigated. A satisfactory form, now widely used, requires some displacement in the y direction but does not substantially alter the schematic in Fig. 7-13. With $w = 0$, the displacements for u and v are given as:

$$u = -\frac{b}{2\pi}\left[\tan^{-1}\frac{y}{x} + \frac{(\lambda + G)}{(\lambda + 2G)}\left(\frac{xy}{x^2 + y^2}\right)\right]$$

$$v = -\frac{b}{2\pi}\left[-\frac{G}{2(\lambda + 2G)}\ell n\frac{x^2 + y^2}{C} + \frac{(\lambda + G)}{(\lambda + 2G)}\left(\frac{y^2}{x^2 + y^2}\right)\right] \quad (7\text{-}21)$$

The term C in the expression for v makes the log term dimensionless and disappears in taking the necessary derivatives.* With the above displacement functions, stress equilibrium is satisfied and the strains based upon these displacements are found from Eqs. (7-6) and (7-7) to be:

$$\gamma_{xy} = \frac{-b}{2\pi(1 - v)}\left[\frac{x(x^2 - y^2)}{(x^2 + y^2)^2}\right]$$

$$e_{xx} = \frac{by}{2\pi(\lambda + 2G)}\left[\frac{Gy^2 + x^2(2\lambda + 3G)}{(x^2 + y^2)^2}\right] \quad (7\text{-}22)$$

$$e_{yy} = \frac{-by}{2\pi(\lambda + 2G)}\left[\frac{-Gy^2 + x^2(2\lambda + G)}{(x^2 + y^2)^2}\right]$$

$$e_{zz} = \gamma_{xz} = \gamma_{zy} = 0$$

The stresses, from Eq. (7-11) are:

$$\tau_{xy} = \frac{-Gb}{2\pi(1 - v)}\left[\frac{x(x^2 - y^2)}{(x^2 + y^2)^2}\right]$$

$$\sigma_x = \frac{Gb}{2\pi(1 - v)}\left[\frac{y(3x^2 + y^2)}{(x^2 + y^2)^2}\right]$$

$$\sigma_y = \frac{-Gb}{2\pi(1 - v)}\left[\frac{y(x^2 - y^2)}{(x^2 + y^2)^2}\right] \quad (7\text{-}23)$$

$$\dagger\sigma_z = \frac{Gbvy}{\pi(1 - v)(x^2 + y^2)}$$

$$\tau_{xz} = \tau_{zy} = 0$$

Because easy glide of edge dislocations is restricted to a slip plane such as that shown in Fig. 7-3(a), the stress of greatest interest and use is that given as τ_{xy} in Eq. (7-23). This will become more evident when the force reactions on dislocations are introduced. Note too that the complete derivations lead-

---

*C is sometimes taken as b^2.

†Since this is a case of plane strain, $\sigma_z = v(\sigma_x + \sigma_y)$.

ing to Eqs. (7-22) and (7-23) using Eq. (7-21) are not included. The interested reader can pursue them independently. In Appendix 7-C a summary of the stress and strain relations is given for both screw and edge dislocations.

7-6 SHEAR STRAIN CAUSED BY DISLOCATION MOVEMENT

Figure 7-14 portrays a crystal of dimensions w, h, and d containing a single dislocation, the line being described by the often-used symbol $\perp$. Plane $ABCD$ is the plane on which slip, shear, or glide is most likely to occur. Note that the dislocation line traverses the full crystal. Assume that this line was originally positioned at the left-hand edge of the grain and that under the influence of an externally applied shear stress, τ, it glides across the full width w until it exits at the right-hand face and causes a small slip step to result. The removal of the dislocation by this process would leave the grain in perfect structural configuration and the induced shear strain would be related to the Burgers vector and crystal height as shown in Fig. 7-15. Since

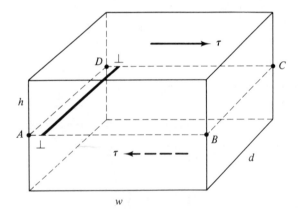

Figure 7-14 Shear stress applied to a crystal containing one edge dislocation.

$b \ll h$, the shear strain caused by the movement of this single dislocation across the full width would be:

$$\gamma_{xy} = \frac{b}{h} \tag{7-24}$$

where the Burgers vector b represents the slip caused by this process. If we now generalize to n dislocations on the same or parallel planes and assume that all are capable of similar movement due to τ, then the induced shear strain would be on the order of:

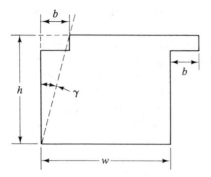

Figure 7-15 Slip caused by traversal of one edge dislocation across the full crystal width.

$$\gamma_{xy} = \frac{nb}{h} \qquad (7\text{-}25)$$

Finally, since many such dislocations, due to original location, might traverse only a portion of the grain, we introduce the idea of an average glide distance; this may be indicated as g/w where g is the average glide distance. Note, for instance, if in Fig. 7-14 the dislocation was initially positioned at the center of the crystal, then g is $\frac{1}{2}w$ instead of w, and the resulting shear strain would be $b/2h$ instead of b/h. The usual expression for illustrating all of these factors is:

$$\gamma_{xy} = \left(\frac{nb}{h}\right)\left(\frac{g}{w}\right) \qquad (7\text{-}26)^*$$

where, b is the Burgers vector, g is the average glide distance and $n/(hw)$ is the dislocation *density*. Note that only those dislocations capable of *easy glide* are to be considered in this context. For many annealed metals, the actual dislocation density is on the order of 10^8 to 10^{10} per m², while highly cold-worked metals indicate densities on the order of 10^{14} to 10^{16} per m². It is common practice to consider such densities in terms of the *area* of grains such as indicated by the product hw above. As will be explained later, and as indicated in Chap. 5, cold-working leads to a greater resistance to slip, so a higher applied stress must be induced to cause increased deformation. The reader may see an anomaly at this point since the weaker, annealed structure contains far fewer dislocations than does the stronger, cold-worked structure. Equation (7-26) might imply that a given applied stress should cause a greater shear strain in the cold-worked material. This is definitely not observed in practice and a tentative answer for now is associated with differences in the average glide distance, g. It is not the total *number* of dislocations

*This also appears as $\gamma_{xy} = nb(A_s/v)$ where A_s is the area swept per dislocation and v is the volume of the crystal.

that is critical in this context; it is the number capable of easy glide that is important.

As a final comment about Eq. (7-26), it finds no direct application in macroscopic design but is significant nonetheless. Since the Burgers vector is basically invariant for a given metal, it does not lend itself to modification. Crystal dimensions can be altered to some extent by combinations of mechanical and thermal treatments to produce different grain sizes. There are, however, a number of other means by which the glide distance can be affected and, together with those of cold-working and altering of grain size, comprise what are commonly called strengthening mechanisms. Regardless of the mechanism employed, a reduction of g is the fundamental reason why strength is affected!

Example 7-2.

With reference to Fig. 7-14, consider this as a crystal where $w = 2$ cm, $h = 1.5$ cm and $d = 2.5$ cm. The metal is made up of atoms whose radius is 1.25 Angstroms and a number of edge dislocations that lie on planes parallel to $ABCD$. Under the applied shear stress, τ, all dislocations move toward the right in that figure. If the *average* glide distance is 0.5 microns, find the number of moving dislocations required to produce a shear strain of 0.05. Use SI units in arriving at the solution.

Solution.

From Eq. (7-26) and the notation beneath that equation:

$$\gamma = 0.05$$
$$b = 2r = 2.5 \text{ Angstroms} = 2.5 \times 10^{-8} \text{ cm} = 0.25 \text{ nm}$$
$$h = 1.5 \text{ cm} = 0.015 \text{ m}$$
$$w = 2 \text{ cm} = 0.02 \text{ m}$$
$$g = 0.5 \text{ microns} = 0.5 \times 10^{-6} \text{ m} = 0.5 \ \mu\text{m}$$
$$n = (\gamma h w)/(g b)$$
$$= [(0.05)(0.015)(0.02)\text{m}^2]/[(0.5)(0.25) \ \mu\text{m}\cdot\text{nm})]$$
$$= [(0.05)(0.015)(0.02)]/[(0.5)(0.25 \times 10^{-15})] = 12 \times 10^{10}$$

dislocations.

7-7 STRAIN ENERGY CAUSED BY DISLOCATIONS

SCREW DISLOCATION

Referring back to the model in Fig. 7-3(b), recall that to produce a screw dislocation demanded the disruption of atoms from their equilibrium sites; this in turn required that elastic energy be introduced by the application of the external force, F. Thus, the presence of actual dislocations must increase the strain energy of a crystal because of atomic disruption from equilibrium positions. Again with regard to Fig. 7-3(b), the larger the grain size, the more elastic energy must be introduced to form a dislocation since a greater number of atoms are displaced from equilibrium. Reference to the block model indicates that work is required to form a dislocation, the work being directly related to the increase in strain energy of the crystal. In Chap. 3, Eq. (3-11) was developed to express work as a function of induced elastic strain energy. Since the application of force F in Fig. 7-3(b) is directly related to the shear stress needed to create the dislocation, the only stress of importance is that of shear as given by Eq. (7-20). For our purposes, therefore, the strain energy caused by the presence of a screw dislocation may be expressed by:

$$dW = \tfrac{1}{2}\tau_{z\theta}\gamma_{z\theta}\,dV \tag{7-27}$$

where dV is the differential volume of concern. If, for simplicity, the grain in which the dislocation energy is to be determined is considered to be a sphere, then the total energy involved will depend upon the outer radius, r_o, of the grain. As mentioned earlier, Eq. (7-20) predicts infinite stresses and strains as the distance r, from the dislocation core approaches zero. To avoid this problem, it is now common practice to consider that the core radius, r_c, is about five times the Burgers vector (i.e., $r_c = 5b$). If the grain is bounded by these radial limits, then the strain energy per *unit length* of dislocation line may be found from

$$W_s = \int_{r_c}^{r_o} \frac{1}{2}\tau_{z\theta}\gamma_{z\theta}\,dV = \frac{1}{2}\int_{r_c}^{r_o} \left(\frac{-Gb}{2\pi r}\right)\left(\frac{-b}{2\pi r}\right)(2\pi r\,dr) \tag{7-28}$$

The solution is:

$$W_s = \frac{Gb^2}{4\pi}\ell n\,\frac{r_o}{r_c} \tag{7-29}$$

where W_s is the strain energy *per unit length* of a screw dislocation line, and G and b are the shear modulus and Burgers vector, respectively.

EDGE DISLOCATION

Figure 7-3(a) indicates that the elastic energy induced by an edge dislocation is primarily related to the shear stress parallel to the slip plane. Using the approach that led to Eq. (7-29) gives the result:

$$W_e = \frac{Gb^2}{4\pi(1-v)}\ell n\left(\frac{r_o}{r_c}\right) \tag{7-30}*$$

*In Eqs. (7-29) and (7-30) the term $\ell n\,[(r_o/r_c) - 1]$ is often given. Cottrell [7] gives this more refined analysis. In this book, the simpler form is used.

where W_e is the strain energy per unit length of an edge dislocation line. Note that due to the $(1 - v)$ term, the edge part of a mixed dislocation causes more energy per unit length than does the screw portion. Thus as a mixed dislocation is induced to move, the edge portion tends to disappear more quickly than the screw portion as this leads to a faster lowering of the total energy. The strain energy within the core of the dislocation, from the line itself to a radius of $5b$, has been estimated to be on the order of 10 percent of the total strain energy, so the use of a cutoff radius such as r_c does not cause any serious errors in the foregoing derivations.

Example 7-3.

In a grain of metal whose radius is 100,000 times its Burgers vector, whose shear modulus is 26 GPa and whose atomic radius is 1.25 Angstroms lies a single screw dislocation. Determine the strain energy per unit length of dislocation line caused by the existence of the dislocation. If Poisson's ratio for this metal is 0.3, how much energy would be induced if this were an edge dislocation?

Solution.

Using Eq. (7-29):

$$W_s = \frac{Gb^2}{4\pi} \ln\left(\frac{r_o}{r_c}\right)$$

If r_c is taken as $5b$, and $b = 2.5$ Angstroms $= 0.25$ nm

$$W_s = \frac{26 \text{ GPa}(0.25)^2 \times 10^{-18} \text{ m}^2}{4\pi} \ln(20,000)$$

$$= \frac{26 \times 10^9 \times (0.25)^2 \times 10^{-18} \times 9.9 \text{ N}}{4\pi}$$

$$= 1.28 \times 10^{-9} \text{ N}$$

Using Eq. (7-30):

$$W_e = \frac{W_s}{(1 - v)} = 1.83 \times 10^{-9} \text{ N}$$

7-8 FORCE ACTING ON A DISLOCATION DUE TO EXTERNALLY APPLIED STRESS

Figure 7-16 portrays a grain containing a single-edge dislocation and an applied shear stress, shown as τ, which acts upon an area wd. Due to τ, the dislocation can move a total distance w whereas the shear force would displace a distance equal to the Burgers vector to cause an amount of slip equal to b. If it is assumed that the amount of work done by the displacement of the shear force is equivalent to the work done to move the dislocation across the full width of the crystal, the introduction of a *fictitious* force acting upon the dislocation line provides a useful concept. This force, shown as f_e in Fig. 7-16, is the force per *unit length* of the line that provides the amount

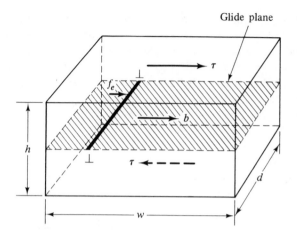

Figure 7-16 Force acting upon an edge dislocation line due to the application of an external shear stress.

of internal work equal to the external work due to τ. The external work

$$W_\tau = \tau \, (wd)b$$

while the internal work

$$W_f = f_e(d)w \tag{7-31}$$

If W_τ equals W_f,

$$f_e = \tau b \tag{7-32}$$

The physical meaning of Eq. (7-32) is that as τ is applied to the crystal, the force f_e is assumed to act on the dislocation line and as τ causes a slip of b, the force, f_e, causes the dislocation to traverse the distance w. Note that for an edge dislocation, f is parallel to the slip plane, acts perpendicular to the dislocation line and is coincident with the direction of motion of the line.

Figure 7-17 is applicable for a screw dislocation. Recall from Fig. 7-3(b) that the creation of such a dislocation required the application of a shear stress as indicated in Fig. 7-17. Under this external stress, τ, the dislocation

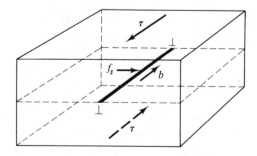

Figure 7-17 Same as Fig. 7-16 considering a screw dislocation.

again moves in a direction perpendicular to the line itself; thus, the fictitious force f_s must also act in the direction of motion. This is often difficult to visualize and it is again noted that reference to the block model provides the most obvious physical rationalization. Since the relationship between τ and f_s is developed from exactly the same arguments just given, the final result is identical to Eq. (7-32). Thus, it is no longer necessary to employ subscripts, so regardless of the type of dislocation involved,

$$f = \tau b \qquad (7\text{-}33)$$

This is sometimes called the Peach-Koehler Equation [10] and gives the force acting per unit length of dislocation line as a function of the external shear stress and Burgers vector. One caution should be noted. With the simple models used in Figs. 7-16 and 7-17, no systematic procedure has been introduced to permit a determination of the actual direction of *motion* of these dislocation lines. This is a deliberate omission since the problems of greatest concern in this chapter do not demand this knowledge. If the reader is interested in pursuing this further, consult Weertman and Weertman[9].

7-9   FORCES DUE TO INTERACTION OF DISLOCATIONS

For simplicity, the developments that follow are restricted to dislocations having *parallel* lines. In addition, edge dislocations will carry the further restriction of lying on the same or *parallel* planes.

SCREW DISLOCATIONS

Suppose in Fig. 7-3(b) a second dislocation having a line parallel and close to LL was introduced in the same way that caused the first dislocation. If this second dislocation has the same *sign* as the first then a greater disruption of atomic alignment results. The shear stress required to produce this new dislocation is felt by the original one as if it were an external shear stress, thus the force it exerts upon the initial dislocation can be considered as being identical to that given in Eq. (7-33). If the second dislocation had the opposite sign from the first, the tendency would be to lower the atomic disruption overall; the second one would tend to remove or annihilate the first. From these physical observations, two screw dislocations of the same sign cause forces to act upon the two lines in a repulsive sense; this leads to greater disruption of the crystal structure (a higher energy level). With two screw dislocations of unlike signs there is mutual attraction which lowers the energy within the crystal. The shear stress caused by a screw dislocation is given by Eq. (7-20) and in combination with Eq. (7-33) the result is:

$$f = \tau_{z\theta}(b) = -\frac{Gb^2}{2\pi r} \qquad (7\text{-}34)^*$$

There is no need to become overly concerned about the sign in Eq. (7-34); it will depend upon the direction of the Burgers vectors of the two parallel screw dislocations which themselves depend upon the choice of the positive sense. Because of the convention used in this book, the negative sign implies repulsion. All one has to remember is that the force f will cause repulsion of like dislocations and attraction of unlike dislocations. Also implicit here is that the magnitude of the two Burgers vectors is identical.† Finally, it should be realized that each dislocation exerts the same force on the other and this force is only a function of the *radial distance* between the dislocations.

Example 7-4.

Two straight and parallel screw dislocations lie in a grain of chromium where the grain diameter is on the order of 30×10^{-4} cm. The dislocations are of opposite sign and are separated by a distance of 1000 Angstroms. Determine the magnitude of the *total* force exerted by each dislocation; will it be repulsive or attractive?

Solution.

From Eq. (7-34) the magnitude of the force per unit length of line is given by:

$$f = \frac{-Gb^2}{2\pi r}$$

and because the dislocations are of unlike signs they will tend to attract each other. The shear modulus of chromium is on the order of 100 GN/m² while the Burgers vector is 2.5 Angstroms or 0.25 nm.

$$f = \frac{\left(100\,\frac{\text{GN}}{\text{m}^2}\right)(0.25)^2(10^{-18})\,\text{m}^2}{2\pi(1000)(10^{-8})(10^{-2})\,\text{m}} = \frac{0.0625}{6.28}\,\frac{\text{N}}{\text{m}} = 10^{-2}\,\frac{\text{N}}{\text{m}}$$

so, the total force

$$F = fl = 10^{-2}\,\frac{\text{N}}{\text{m}}\,(30 \times 10^{-6}\,\text{m})$$

$$F = 30 \times 10^{-8}\,\text{N} = 0.3\ \mu\text{N}$$

EDGE DISLOCATIONS

Figure 7-18 shows two parallel edge dislocations, A and B, of the same sign positioned on parallel planes. Such an arrangement may be envisioned with the aid of Fig. 7-3(a) where a second cut and displacement is performed. As compared with the explanation regarding screw dislocations of like or

---

*This could be expressed in terms of x and y but the form shown is most useful.

†It is more correct to use $b_1 \cdot b_2$ instead of b^2 to account for possible differences in the Burgers vectors of the two disolcations. In this text it is *always* assumed that $b_1 = b_2$ in magnitude; the signs, however, may differ.

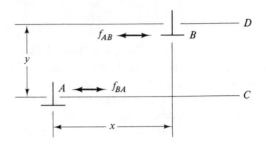

Figure 7-18 Forces between parallel edge dislocations on parallel glide planes.

unlike signs, an equivalent explanation for edge dislocations cannot be given on the basis of physical intuition alone regarding the repulsion or attraction of these two edge dislocations. One must accept that if the assumptions which led to Eq. (7-23) were reasonable, then any findings that come from the use of those equations are acceptable.

For the moment, consider that the force reactions between the two dislocations, whether attractive or repulsive, will tend to cause motion due to easy glide on the two parallel slip planes, C and D. Although other forces are felt by each dislocation, since normal *stresses* are indicated by Eq. (7-23), it is the stress τ_{xy} that is of primary importance. Forces due to the normal stresses would attempt to cause dislocation motion by severing many atomic bonds simultaneously; this gets back to the basic reason why τ_{max} is so much larger than the observed stresses needed to induce slip. For this reason, the influence of these normal stresses is minimal.* It is again useful to consider that the important forces felt by the lines of dislocations A and B arise because the shear stress, τ_{xy}, tends to produce such forces. As contrasted with the findings where parallel screw dislocations were discussed, the influential stress, τ_{xy}, acts parallel to the planes of easy glide, C and D, and produces a force per unit length that acts mutually upon the two dislocations. This force is perpendicular to the dislocation lines. The direction of movement of each line is, as mentioned earlier, at right angles to the line itself. By combining Eqs. (7-23) and (7-33),

$$f = \tau_{xy}(b) = -\frac{Gb^2}{2\pi(1-v)}\left[\frac{x(x^2-y^2)}{(x^2+y^2)^2}\right] \qquad (7\text{-}35)$$

For convenience in displaying this result graphically, the coefficient, which is a function of material property values, is given by:

$$C = \frac{Gb^2}{2\pi(1-v)} \qquad (7\text{-}36)$$

*This is called climb; see Sec. 7-11.

Multiplying this equation by y/y and rearranging gives:

$$\frac{fy}{C} = -xy\frac{(x^2 - y^2)}{(x^2 + y^2)^2} \tag{7-37}$$

It is common practice to plot $(fy)/C$ versus x where x is given in multiples of y as shown in Fig. 7-19. This plot is due to Cottrell [7]. The solid line is used for dislocations of like signs, whereas the dotted line, an exact mirror image, pertains for unlike signs. For either case, a positive value of $(fy)/C$ implies attraction while a negative value indicates repulsion as shown on that figure.* Several important observations are:

1. When $x = 0$ and dislocations of like signs are involved, a position of stable equilibrium is indicated since any tendency for motion

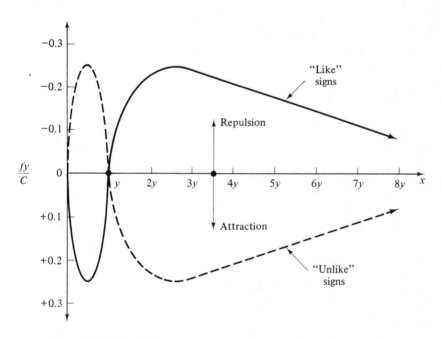

Figure 7-19 Plot of forces between parallel edge dislocations on parallel glide planes. Solid curve is for dislocations of the same sign and dotted curve pertains to unlike signs. In both cases, the forces are those parallel to the glide planes.

*The signs on the ordinate of Fig. 7-19 are opposite to those usually shown. This results from different conventions used to define a positive and negative dislocation. In this text, Eq. (7-35) expresses the result for dislocations of the same sign *and* magnitude! If dislocations were of unlike *signs*, then a positive sign would be indicated in Eq. (7-35). With this in mind, the relative values of x and y then provide consistency with Fig. 7-19.

would be opposed by the attraction indicated. Dislocations of unlike signs are unstable at $x = 0$, since repulsive forces occur when $0 < x < y$.

2. When $x = y$, a condition of quasi-equilibrium prevails for dislocations of like signs while dislocations of unlike signs are in a stable condition.

3. If the dislocations lie on the same glide plane, Fig. 7-19 *cannot* be used. One must resort to Eq. (7-35) setting $y = 0$. The magnitude of the force is, then,

$$f = \frac{C}{x} \qquad (7\text{-}38)$$

where dislocations of like signs induce repulsive forces whose magnitude is a function of the distance x. If of unlike signs, the dislocations attract and annihilate each other since the two extra half-planes that lie above and below the glide plane would combine to form a continuous alignment. Figure 7-3(a) provides a physical feel in this regard.

To summarize the important aspects of this section on edge dislocations:

1. For two having the same sign and on parallel planes:
    (a)   Repulsion if $x > y$.
    (b)   Attraction if $x < y$.
2. For two having opposite signs and on parallel planes:
    (a)   Attraction if $x > y$.
    (b)   Repulsion if $x < y$.
3. For two on the same plane:
    (a)   Attraction if of opposite signs.
    (b)   Repulsion if of the same sign.

For situations one and two above, Eq. (7-35) can always be used to give an exact answer whereas Fig. 7-19 provides a quick approximation, the accuracy of which depends upon the exactness with which the plot is interpreted. The minimum stress required to cause dislocation movement is usually called the Peierls stress and is attributed to the work of Peierls [11] and Nabarro [12].

Example 7-5.

Two edge dislocations of unlike signs are spaced horizontally by a distance of 100 nm and they lie on parallel planes separated by a distance of 25 nm.
(a)   Will they tend to attract or repel each other?
(b)   What is the magnitude of the *initial* attractive or repulsive force per unit length of dislocation line?

Solution.

From Fig. 7-19, $x = 100$ nm $= 4y$, and the use of the dotted curve at $x = 4y$ indicates an attractive force where $(fy)/C$ is read off as approximately $+0.21$.
Using Eq. (7-37):

$$\frac{fy}{C} = +\frac{4y(y)(16y^2 - y^2)}{(16y^2 + y^2)^2} = +\frac{4(15)}{(17)^2} = +0.208$$

To determine the actual force per unit length would require a calculation for the value of C, based upon material properties, using Eq. (7-36). Since the purpose of this example was to demonstrate the use of Fig. 7-19 and its equivalence with Eq. (7-37), there is no need to proceed with an actual calculation of C.

Example 7-6.

Under an external shear stress, an edge dislocation moves in a grain of copper until it is stopped along its entire length by some type of barrier. A second dislocation of the same sign lies on the same slip plane and moves towards the pinned dislocation. If the applied shear stress is 1 MPa and no other factors are considered, what will be the final equilibrium spacing between the two dislocations?

Solution.

Since these dislocations are of like signs and lie on the same plane, Eq. (7-35) degenerates to Eq. (7-38) since $y = 0$ and the pinned dislocation tends to repel the second one with a force:

$$f = \tau_{xy}b$$

The applied shear stress, τ_a, produces a force per unit length of $f = \tau_a b$ on the second dislocation in a direction *opposite* the force exerted by the pinned dislocation, thus

$$f_1 = \tau_{xy}\, b \longrightarrow \perp \longleftarrow f_2 = \tau_a b$$

and equilibrium occurs when the net force is zero or when

$$\tau_a = \tau_{xy} \quad \text{or} \quad \tau_a = \frac{Gb}{2\pi(1 - v)}\left(\frac{1}{x}\right)$$

where x is the equilibrium spacing. Thus

$$x = \frac{Gb}{2\pi\tau_a(1 - v)}$$

For copper, $G = 42$ GN/m^2, $b = 0.256$ nm and $v = 0.35$

$$x = \frac{(42)(0.256)}{2\pi(0.65)(10^6) \text{ N/m}^2}(10^9)\frac{\text{N}}{\text{m}^2}(10^{-9}) \text{ m}$$

$$= \frac{42(0.256)(10^{-6}) \text{ m}}{2\pi(0.65)} = 2.63 \times 10^{-6} \text{ m} = 2.63 \ \mu\text{m}$$

Note that the important material properties must be found from other sources. This is intentional. In real life situations, engineers must find the important data of concern and students should be exposed to this necessity at an early stage of their academic pursuits.

7-10 GENERATION OF DISLOCATIONS

As mentioned earlier, a large increase in dislocation density occurs when annealed metals are subjected to increasing amounts of cold-working. It is of interest to inquire as to how this result might occur. The most commonly used model in this regard is called a Frank-Read [13] source or generator. The sequence of events portrayed in Fig. 7-20 shows how dislocations are

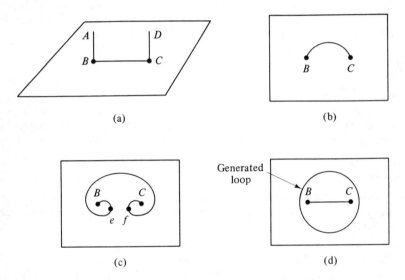

(a)

(b)

(c)

(d)

Figure 7-20 Frank-Read model for the generation of dislocations.

generated by this model. The dislocation line is $ABCD$ and only the portion BC lies in the slip plane; points B and C are pinned in position, AB and CD being immobile. Under an applied shear stress, BC begins to bow until it forms a semicircle. Further deformation leads to an unstable situation where the dislocation takes the configuration shown in Fig. 7-20(c). When the points e and f coincide, those parts of the loops are of opposite sign and annihilation results. A full loop is then formed and the portion BC restored to its initial position. By continually repeating this process, additional dislocation loops may be produced until the ever expanding loops meet adequate resistance to shut off further generation. Figure 7-21 by Dash [14] shows an exceptional photograph that provides definite support for this proposed model.

As BC tends to bow under the influence of the stress τ, an increase in l occurs. To stretch this line demands an increase in its stored energy, much like stretching a rubber band. Considering the force analysis in Fig. 7-22 and using Eq. (7-33), the *total* force acting on the line is:

$$fl = \tau bl = F \qquad (7-39)$$

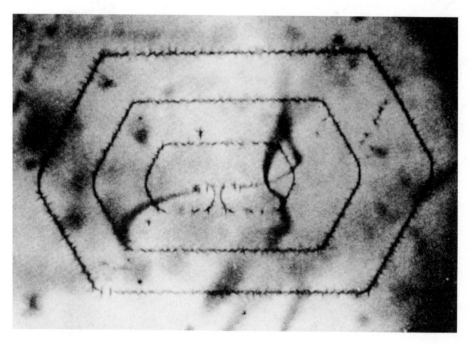

Figure 7-21 Dislocation loops generated by a Frank-Read source in a crystal of silicon (from Dash [14] with the kind permission of John C. Fisher).

If the increase in strain energy is expressed by Eq. (7-30), then the force balance, shown in Fig. 7-22, is:

$$\tau bl = 2\,W_e \sin \theta \tag{7-40}$$

The instability condition occurs when $\theta = 90°$, so the applied stress required to bow BC into a semicircle is:

$$\tau = \frac{2W_e}{bl} \tag{7-41}$$

W_e is given by Eq. (7-30), so,

$$\tau = \frac{2}{lb}\left[\frac{Gb^2}{4\pi(1-v)}\,\ell n\left(\frac{r_o}{r_c}\right)\right] \tag{7-42}$$

or

$$\tau = \frac{2Gb}{l}K \quad \text{where} \quad K = \frac{1}{4\pi(1-v)}\,\ell n\left(\frac{r_o}{r_c}\right) \tag{7-43}$$

Considering v to be on the order of 0.3 and r_c to be $5b$, a variation in r_o from

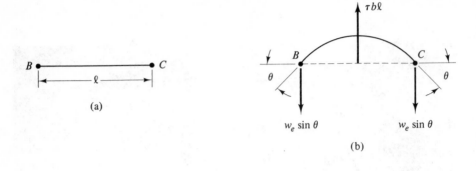

(a)

(b)

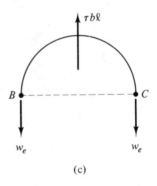

(c)

Figure 7-22 Determination of the stress required to start dislocation generation using a force balance on the pinned dislocation line.

$50b$ to $5 \times 10^5 b$ causes K to vary from about 0.26 to 1.30. Because of the ℓn function, K varies very little over a wide range of r_o. As a consequence and for simplicity, an average value for K is often taken as $\frac{1}{2}$. Using this value:

$$\tau_F \approx \frac{Gb}{l} \qquad (7\text{-}44)^*$$

where τ_F is the applied shear stress required to start the generation of a Frank-Read source.†

*$\tau = (2Gb)/l$ is also given; this depends upon the value of K used which, itself, depends upon the arbitrary choice of r_o. Note too that if Eq. (7-29) is used to describe the strain energy and, the $(1 - v)$ term ignored, the variation of K is from 0.18 to 0.92 for the range of r_o used prior to Eq. (7-44). This variation causes minimal changes and the result given by Eq. (7-44) is still appropriate.

†The cross-slip (see Sec. 7-11) of screw dislocations can also lead to a Frank-Read source. References [1] and [2] discuss this.

Example 7-7.

Suppose a metal possesses the following properties:
(a) $E = 15 \times 10^6$ psi
(b) $v = 0.3$
(c) $r = 1.28$ Å

A dislocation is pinned between two obstacles that lie 200 Å apart whereas the grain itself is on the order of a micron in diameter. What is the magnitude of the applied stress needed to start the generation of dislocations according to the Frank-Read model?

Solution.

Using Eq. (7-44) as a reasonable approximation:

$$\tau_F = \frac{Gb}{l} \quad \text{where} \quad G = \frac{E}{2(1+v)} = 5.8 \times 10^6 \text{ psi}$$

$$b = 2.56 \text{ Å}, \quad l = 200 \text{ Å}$$

$$\tau_F = \frac{5.8(10)^6(2.56)}{200} \text{ psi} = 74,240 \text{ psi}$$

Consider the alternative approach using Eq. (7-42) where

$$r_o = \tfrac{1}{2} \text{ micron} \quad \text{and} \quad r_c = 5b = 12.8 \text{ Å}$$

$$\tau_F = \frac{2Gb}{4\pi l(1-v)} \ell n \left(\frac{r_o}{r_c}\right)$$

$$\tau_F = \frac{2(5.8)(10^6)(2.56)}{4\pi(200)(0.7)} \ell n \left(\frac{5000}{12.8}\right)$$

$$= 1.69(10^4)(5.97) = 100,890 \text{ psi}$$

If the $(1-v)$ term is omitted, then $\tau_F = 70,623$ psi.

7-11 CLIMB AND CROSS SLIP

Although no attempt will be made in regard to an extended discussion, a brief description of the climb and cross slip of dislocations seems appropriate. It is possible for edge dislocations to move *perpendicular* to their slip plane by the process of climb. In general, two factors are required if climb is to occur. The first is the existence of vacancies. These are nothing more than the absence of an atom at a site normally occupied. The second factor is a high enough temperature, since climb is a diffusion type of process. Figure 7-23 shows a schematic of the climb process where the edge dislocation has moved vertically by one atomic plane. This is an important mechanism since it introduces an additional freedom for dislocation motion if the initial slip

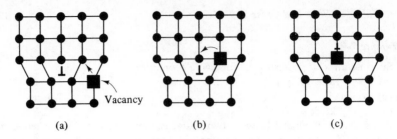

(a)   (b)   (c)

Figure 7-23 Illustration of dislocation climb.

plane impedes easy glide of the edge dislocation. To discuss the detailed ramifications of climb is beyond the intent of this text.

Cross-slip is a process whereby a screw dislocation encounters some obstacle on a current slip plane and simply changes slip planes by crossing to an adjacent one. Recall from the discussion connected with Eq. (7-20) that a screw dislocation was not restricted to a single slip plane as is the usual case for an edge dislocation. Figure 7-24 depicts the cross-slip process. Other texts [1, 2] provide a fuller discussion on climb and crossslip.

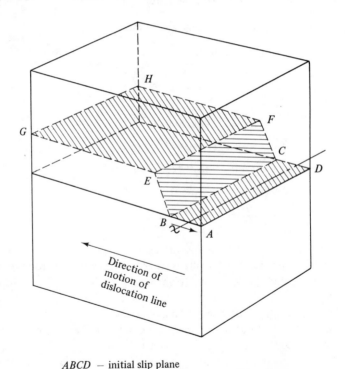

$ABCD$ — initial slip plane
$EFGH$ — new slip plane after a screw dislocation has "cross slipped."

Figure 7-24 Illustration of cross-slip.

7-12   QUALITATIVE EXPLANATIONS OF MACROSCOPIC BEHAVIOR

Certain qualitative explanations of the observed macroscopic behavior of crystalline solids can now be given using the concepts of dislocation theory. In all instances, metallic solids are implied.

EFFECT OF GRAIN SIZE

Figure 7-25 illustrates single grains for a given metal where the only intended difference is the average size of individual grains. Considering a single dislocation in each grain, it is apparent that the average glide distance is smaller in the finer grain. Thus for a given applied shear stress, a lower shear strain will result in the smaller grain and that structure will be stronger. Refining grain size thereby provides a strengthening mechanism.*

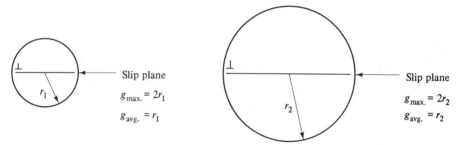

Figure 7-25 Effect of grain size upon average glide distance.

COLD-WORKING

Due to the generation of dislocations from cold-working, a pile-up can occur as shown in Fig. 7-26 where the grain boundary acts as a barrier. The distance from the Frank-Read source to the grain boundary limits the total number of dislocations that can be generated since the force interactions between all dislocations leads to a final balancing and the source ceases to operate. Although the total number of dislocations has increased, motion by easy glide has decreased and higher applied stresses would be required to produce further movement. In this way, strength has been increased. Dislocations on nonparallel slip planes may also intersect and produce what are called jogs. A futher discussion of such effects will not be pursued here but it is important to realize that such results reduce the tendency of easy glide and lead to increased strength.

*This is correct when temperatures are low enough to cause negligible effects due to diffusion.

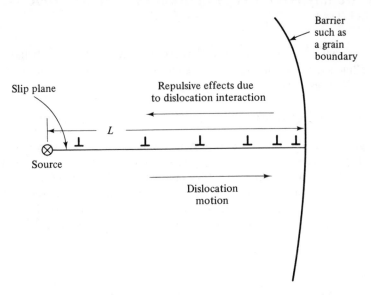

Figure 7-26 Dislocation pile-up at a grain boundary. The total number of dislocations in a pile-up is approximated by $n = (2\sigma L)/(bG)$, where L is the distance shown, n is the number in the pile-up and σ, b, and G are the applied stress, Burgers vector, and shear modulus, respectively. Cottrell [7] provides a more accurate analysis.

BAUSCHINGER EFFECT

It has long been observed that if a metal is plastically deformed say in tension, then the direction of loading reversed by compression, the stress required to cause plastic deformation is lower in compression than the level that would be required to cause further deformation under tensile application. This is called the *Bauschinger effect*. One explanation can be given using Fig. 7-26. Every dislocation is influenced by the interactive forces of all others and, due to the cumulative repulsive effects, the spacing between dislocations increases away from the barrier. Suppose the pile-up was due to tensile stress application. To force the dislocations closer together requires an increase in the tensile stress. However, if the *direction* of applied stress is reversed, then the repulsive forces due to dislocation interaction will tend to assist rather than oppose dislocation motion. Because of these back stresses, the magnitude of the applied compressive stress will be lower than that needed in tension if further plastic deformation is to occur. For an explanation at the macroscopic level, see Polakowski and Ripling [2].

RECOVERY*

If cold-worked metals are subjected to ever-increasing levels of temperature, a complex sequence of events can occur. This process itself is usually described by the term annealing which is then further subdivided into regions called recovery, recrystallization, and grain growth. Recovery occurs at the lower end of the temperature range, grain growth occurs at the higher levels and recrystallization at intermediate levels.† The description of events leading to recrystallization and grain growth is quite similar for all metals whereas the changes that occur during recovery show subtle differences regarding various metallic alloys.

As an example, with metals that undergo phase transformations, consider a typical steel that transforms from austenite to martensite under rapid cooling. If such a metal were reheated to fairly low temperatures, say 150°C, there would be no obvious change in macroscopic hardness but a decided increase in tensile strength could follow. This is usually called *stress relieving*; it is caused by an internal relief of stresses that are introduced originally by a nonuniform combination of the effects of phase transformation and thermal expansion and contraction. At somewhat higher temperature levels, the effect of recovery then sees a drop in *both* hardness and strength, thus, with metallic alloys such as steels, recovery causes a softening or weakening to occur.

This explanation however, is not fully applicable where single-phase alloys such as alpha brasses are involved. If initially cold-worked specimens are reheated to temperatures higher than room temperature but less than the temperature to start recrystallization, a modest but definite increase in hardness would be noted. As the reheating temperature is increased further, the structure begins to recrystallize and an obvious decrease in hardness follows. For many years, the range of temperatures below the onset of recrystallization has been termed the recovery region and all that is attempted here is a plausible explanation of why the hardness of certain signle-phase metals displays an *increase* at these lower temperature levels. Figure 7-27(a) shows a number of like dislocations on parallel slip planes after straining; note the randomness of positions. At temperature levels typical of recovery, an adequate amount of thermal input can cause movement of the dislocations such that the alignment shown in Fig. 7-27(b) results. In effect, the dislocations tend to stack in a vertical alignment. Reference to Fig. 7-19 shows that this is a condition of stable equilibrium as far as the interaction forces are involved. The arrangement in Fig. 7-27(b) is called polygonization and the

---

*As used here, this word has an entirely different meaning from the definition in Chap. 6.

†These effects are also time-dependent.

vertical alignment forms a type of grain boundary.* Due to the usual influence of grain boundary effects discussed earlier in this section, and since the
vertical alignment is one of stable equilibrium, the increase in strength and
hardness of certain single-phase metals as influenced by recovery is certainly
understandable.

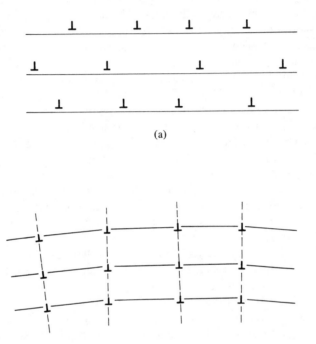

(a)

(b)

Figure 7-27 Generalized positioning of dislocations after cold working (a)
and after polygonization (b).

Fine Particle Strengthening

If particles of a harder phase are present in the matrix of a metallic solid,
they will act to inhibit easy glide of dislocations. Whether they exist as a
result of precipitation effects (one phase at elevated temperatures and two
phases at low temperatures) or simple dispersion (two phases at both tem-

*Such an array of edge dislocations is called a *tilt boundary*, whereas *twist boundary* is
used for screw dislocations. Small angle boundary is often used in either case.

perature levels), the same model serves to explain the end results.* Figure 7-28 portrays the model most often used to describe this strengthening mechanism. In both portions of that figure, the total *amount* of the second

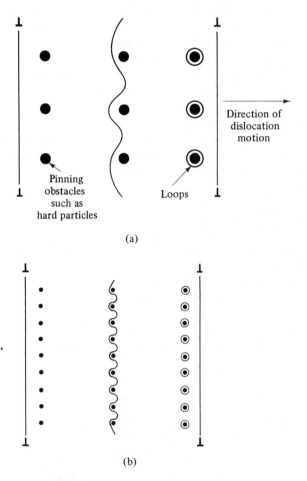

(a)

(b)

Figure 7-28 Model illustrating the strengthening effect due to the presence of fine particles of two different sizes.

*With many copper-aluminum alloys, the formation of a precipitate, from an initially supersaturated single phase, does increase hardness and is called precipitation hardening. With many martensitic structures of steel, also a supersaturated single phase, reheating to produce a precipitate shows a *decrease* in hardness. Perhaps this should be called precipitation softening.

phase is identical, but fewer of the larger particles exist in Fig. 7-28(a). As the dislocation line encounters the row of particles, it is forced to lengthen as it bends around the row. Recalling that the energy of a dislocation is directly related to its length, an increase in the applied stress is demanded. In the sequence of sketches shown, an ever-increasing stress is required for further dislocation motion. One possible eventuality would be the formation of loops around the particles and the eventual reuniting of the dislocation as indicated. As other dislocations approach the particles, the same process may continue. If a greater number of smaller particles exists initially, the smaller curvature and additional increase in the length of the original line simply means that higher stresses are needed to effect the same behavior. This is shown in Fig. 7-28(b). It is of interest to note that when *overaging* of alloys occurs, the large number of initially fine particles have altered to a smaller number of larger particles due to the process of coalescence. A decided drop in strength results and the model discussed above would agree with that observation.

MICROCRACK FORMATION

Several theories, related to different types of solids, have been proposed to explain how microcracks might form but only one is discussed here. In Fig. 7-29, the sketch indicates how a number of dislocations might coalesce to form an empty region much larger than that caused by a single dislocation. In this way a microflaw is introduced and the possible consequences at a macroscopic level are discussed in Chap. 8. Note that flaws or cracks could result for reasons other than those based upon dislocation concepts.

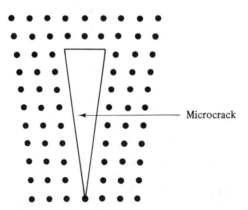

Figure 7-29 Model showing formation of a microcrack due to coalescence of dislocations.

PRONOUNCED YIELD POINT AND STRAIN-AGING

Some metals, such as annealed low-carbon steel, display a pronounced yield point when a sufficient level of stress is applied. One of the more widely accepted explanations of this phenomenon considers the influence of interstitial atoms such as nitrogen or carbon to be most responsible for this behavior. Based upon energy arguments, the small interstitial atoms show a preference to locate themselves in the voids around dislocations. There they act to pin the dislocation against easy glide and result in the demand of a sufficiently high stress if the dislocation is to be pulled away from the pinning effect. When a dislocation escapes from this region it can then move at a lower stress and a saw-tooth behavior is observed on the stress-strain curve until all anchored dislocations have been freed from the interstitial atoms. Uniform changes then occur as further plastic deformation occurs under an increasing stress.* Figure 7-30(a) typifies this behavior. Numerous experiments involving the alteration of carbon and nitrogen, as well as the influence of other elements, appear to reasonably support the above explanation.

It has also been noted that if plastic deformation is interrupted at a point where the induced strain is beyond the initial yield point and the specimen is unloaded then reloaded in a short time, the new stress-strain curve would follow the behavior indicated in Fig. 7-30(b). However, if the reloading is

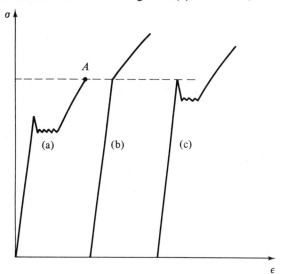

Figure 7-30 Illustration of strain aging.

*The flat saw-tooth region corresponds with nonuniform plastic deformation across the specimen surface. These are called Lüders lines, bands, or strains.

delayed for a substantial time period or if the specimen is subjected to adequate temperature levels, the reloading behavior will display a *return* of a pronounced yield point. This is shown by Fig. 7-30(c), and it should be noted that the new yield stress is higher than the original value because of the cold-working effect. Return of a yield point in this manner is called strain-aging and is held to be caused by the diffusion of interstitial atoms to the vicinity of dislocations; the pinning or anchoring effect is then again introduced.

The explanations in Sec. 7-12 are not exhaustive nor at all complete since other theoretical explanations can be found in the vast literature concerning dislocation theory. What is intended is that the reader gain a degree of insight and some appreciation of the importance of dislocation theory as a reasonable way of explaining macroscopic behavior from microscopic concepts.

REFERENCES

[1] R. W. HERTZBERG, *Deformation and Fracture Mechanics of Engineering Materials* (New York: John Wiley & Sons, Inc., 1976), pp. 49–51, 60–65, 101–26.

[2] N. H. POLAKOWSKI and E. J. RIPLING, *Strength and Structure of Engineering Materials* (Englewood Cliffs, N. J.: Prentice-Hall Inc., 1966), pp. 144–46, 160–61, 177–79, 184–87, 187–88.

[3] F. A. McCLINTOCK and A. S. ARGON, *Mechanical Behavior of Materials* (Reading, Mass.: Addison-Wesley Publishing Co., Inc., 1966), pp. 24, 34, 118–121, 555.

[4] G. I. TAYLOR, "The Mechanism of Plastic Deformation of Crystals, Part I, Theoretical," *Proc. Roy. Soc. (London)*, **A145** (1934), pp. 362–87.

[5] E. OROWAN, "Crystal Plasticity III, on the Mechanism of the Glide Process," *Zeit. Phys.*, **89** (1934), pp. 634–59.

[6] M. POLANYI, "On a Kind of Glide Disturbance that Could Make a Crystal Plastic," *Zeit. Phys.*, **89** (1934), pp. 660–64.

[7] A. H. COTTRELL, *Dislocations and Plastic Flow in Crystals* (Oxford: Clarendon Press, 1953), pp. 9–11, 37–8, 46–9, 104–107.

[8] J. M. BURGERS, "Some Considerations in the Field of Stress Connected with Dislocations in a Regular Crystal Lattice," *Proc. Kon. Ned. Akad. Wet.*, **42** (1939), pp. 378–99.

[9] J. WEERTMAN and J. R. WEERTMAN, *Elementary Dislocation Theory*, (New York: The MacMillan Co., 1964), pp. 18–19, 22–41.

[10] M. PEACH and J. S. KOEHLER, "The Forces Exerted on Dislocations and the Stress Fields Produced by Them," *Phys. Rev.*, **80** (1950), pp. 436–39.

[11] R. PEIERLS, "The Size of a Dislocation," *Proc. Phys. Soc.*, **52** (1940), pp. 34–7.

[12]   F. R. N. NABARRO, "Dislocations in a Single Cubic Lattice," *Proc. Phys. Soc.,* **59** (1947), pp. 256–72.

[13]   F. C. FRANK and W. T. READ, "Multiplication Processes for Slow Moving Dislocations," *Phys. Rev.,* **79** (1950), pp. 722–723.

[14]   W. C. DASH, *Dislocations and Mechanical Properties of Crystals,* ed. J. C. Fisher (New York: John Wiley, and Sons, Inc., 1957), pp. 57–68.

Appendix 7-A
DERIVATION OF EQ. (7-11)

From Eq. (7-8),

$$e_x = \frac{\sigma_x}{E} - \frac{\nu}{E}(\sigma_y + \sigma_z)$$

then,

$$\sigma_x = Ee_x + \nu(\sigma_y + \sigma_z) \tag{7A-1}$$

or, using Eq. (3-5),

$$\sigma_x(1 + \nu) = Ee_x + \nu(\sigma_x + \sigma_y + \sigma_z) = Ee_x + 3\nu\sigma_m \tag{7A-2}$$

With the use of Eq. (3-6), (7A-2) may be expressed by:

$$\sigma_x(1 + \nu) = Ee_x + \frac{\nu \Delta E}{(1 - 2\nu)} \tag{7A-3}$$

or,

$$\sigma_x = \frac{Ee_x}{(1 + \nu)} + \frac{\nu \Delta E}{(1 + \nu)(1 - 2\nu)} = 2Ge_x + \frac{\nu \Delta E}{(1 + \nu)(1 - 2\nu)} \tag{7A-4}$$

Using the Lamé constant, given by Eq. (7-10), (7A-4) becomes:

$$\sigma_x = 2Ge_x + \lambda \Delta = (\lambda + 2G)e_x + \lambda(e_y + e_z) \tag{7A-5}$$

where Δ is defined by Eq. (3-4).

This is identical to the first expression in Eq. (7-11). The other expressions result using an interchange of subscripts.

Appendix 7-B
DERIVATION OF EQ. (7-13)

For the case where all $\partial/(\partial z) = 0$, Eq. (7-12) becomes:

$$\frac{\partial \sigma_x}{\partial x} + \frac{\partial \tau_{xy}}{\partial y} = 0 \tag{7B-1}$$

Consider the first expression in Eq. (7-11) where

$$\sigma_x = (\lambda + 2G)e_x + \lambda e_y + \lambda e_z \tag{7B-2}$$

From Eq. (7-6) (with $\partial/(\partial z) = 0$ and $w = 0$):

$$e_x = \frac{\partial u}{\partial x}, \qquad e_y = \frac{\partial v}{\partial y}, \qquad e_z = 0,$$

$$\gamma_{xy} = \frac{\partial v}{\partial x} + \frac{\partial u}{\partial y}, \qquad \gamma_{yz} = \gamma_{zx} = 0 \tag{7B-3}$$

Differentiating (7B-2) with respect to x gives:

$$\frac{\partial \sigma_x}{\partial x} = (\lambda + 2G)\frac{\partial e_x}{\partial x} + \lambda \frac{\partial e_y}{\partial x} \tag{7B-4}$$

Since $\tau_{xy} = G\gamma_{xy}$

$$\frac{\partial \tau_{xy}}{\partial y} = G\frac{\partial \gamma_{xy}}{\partial y} \tag{7B-5}$$

The relations in (7B-3) are differentiated with respect to x and y as given below:

$$\frac{\partial e_x}{\partial x} = \frac{\partial^2 u}{\partial x^2}, \qquad \frac{\partial e_y}{\partial x} = \frac{\partial^2 v}{\partial y\,\partial x} \qquad \text{and} \qquad \frac{\partial \gamma_{xy}}{\partial y} = \frac{\partial^2 v}{\partial x\,\partial y} + \frac{\partial^2 u}{\partial y^2} \tag{7B-6}$$

Combining (7B-4), (7B-5) and (7B-6) gives:

$$\frac{\partial \sigma_x}{\partial x} = (\lambda + 2G)\frac{\partial^2 u}{\partial x^2} + \lambda \frac{\partial^2 v}{\partial y\,\partial x} \tag{7B-7}$$

and

$$\frac{\partial \tau_{xy}}{\partial y} = G\left(\frac{\partial^2 v}{\partial x\,\partial y} + \frac{\partial^2 u}{\partial y^2}\right) \tag{7B-8}$$

Introducing (7B-7) and (7B-8) into (7B-1) gives:

$$(\lambda + 2G)\frac{\partial^2 u}{\partial x^2} + G\frac{\partial^2 u}{\partial y^2} + (\lambda + G)\frac{\partial^2 v}{\partial x\,\partial y} = 0 \tag{7B-9}$$

This is identical to the first expression in Eq. (7-13). An identical procedure would produce the other two expressions.

Appendix 7-C
SUMMARY OF STRESS AND STRAIN RELATIONS FOR EDGE AND SCREW DISLOCATIONS

STRESSES IN CARTESIAN COORDINATES

Screw Dislocation

$$\tau_{yz} = \frac{-Gb}{2\pi}\frac{x}{(x^2 + y^2)}, \qquad \tau_{zx} = \frac{Gb}{2\pi}\frac{y}{(x^2 + y^2)}$$

$$\sigma_x = \sigma_y = \sigma_z = \tau_{xy} = 0$$

Edge Dislocation $[D = (Gb)/2\pi(1 - v)]$

$$\tau_{xy} = -Dx(x^2 - y^2)/(x^2 + y^2)^2$$
$$\sigma_x = Dy(3x^2 + y^2)/(x^2 + y^2)^2$$
$$\sigma_y = -Dy(x^2 - y^2)/(x^2 + y^2)^2$$
$$\sigma_z = v(\sigma_x + \sigma_y) = 2Dvy/(x^2 + y^2)$$
$$\tau_{xz} = \tau_{yz} = 0$$

STRESSES IN CYLINDRICAL COORDINATES

Screw Dislocation

$$\tau_{r\theta} = \tau_{rz} = \sigma_r = \sigma_\theta = \sigma_z = 0, \tau_{z\theta} = \frac{-Gb}{2\pi r}$$

Edge Dislocation $[D = (Gb)/2(1 - v)]$

$$\sigma_r = \sigma_\theta = D \sin \theta/r, \qquad \sigma_z = 2Dv \sin \theta/r$$
$$\tau_{r\theta} = -D \cos \theta/r, \qquad \tau_{rz} = \tau_{z\theta} = 0$$

STRAINS IN CARTESIAN COORDINATES

Screw Dislocation

$$e_x = e_y = e_z = \gamma_{xy} = 0$$
$$\gamma_{yz} = \frac{-b}{2\pi} \frac{x}{(x^2 + y^2)}, \qquad \gamma_{zx} = \frac{b}{2\pi} \frac{y}{(x^2 + y^2)}$$

Edge Dislocation

$$e_x = \frac{by}{2\pi} \frac{[Gy^2 + x^2(2\lambda + 3G)]}{(\lambda + 2G)(x^2 + y^2)^2}$$
$$e_y = \frac{-by}{2\pi} \frac{[x^2(2\lambda + G) - Gy^2]}{(\lambda + 2G)(x^2 + y^2)^2}$$
$$\gamma_{xy} = \frac{-b}{2\pi(1 - v)} \frac{[x(x^2 - y^2)]}{(x^2 + y^2)^2}$$
$$e_z = \gamma_{xz} = \gamma_{zy} = 0$$

STRAINS IN CYLINDRICAL COORDINATES

Screw Dislocation

$$e_r = e_\theta = e_z = \gamma_{r\theta} = \gamma_{rz} = 0, \qquad \gamma_{z\theta} = \frac{-b}{2\pi r}$$

Edge Dislocation

$$e_\theta = e_r = \frac{b}{4\pi} \frac{(1-2v)}{(1-v)} \frac{\sin\theta}{r}$$

$$\gamma_{r\theta} = \frac{-b}{2\pi(1-v)} \frac{\cos\theta}{r}$$

$$e_z = \gamma_{rz} = \gamma_{z\theta} = 0$$

In terms of λ and G, the strains for an edge dislocation are:

$$e_\theta = e_r = \frac{b}{2\pi} \frac{G}{(\lambda+2G)} \frac{\sin\theta}{r}$$

$$\gamma_{r\theta} = \frac{-b}{\pi} \frac{(\lambda+G)}{(\lambda+2G)} \frac{\cos\theta}{r}$$

These may be developed from the use of $v = \lambda/[2(\lambda + G)]$. Also, to convert from cartesian to cylindrical coordinates, the following relations must be used:

$$u_r = u\cos\theta + v\sin\theta$$

$$u_\theta = -u\sin\theta + v\cos\theta$$

where $x = r\cos\theta$, $y = r\sin\theta$, $r^2 = x^2 + y^2$, $\theta = \tan^{-1} y/x$ and the strains are found from:

$$e_r = \frac{\partial u_r}{\partial r}, \qquad e_\theta = \frac{1}{r}\frac{\partial u_\theta}{\partial\theta} + \frac{u_r}{r}, \qquad e_z = \frac{\partial w}{\partial z}$$

$$\gamma_{r\theta} = \frac{\partial u_\theta}{\partial r} - \frac{u_\theta}{r} + \frac{1}{r}\frac{\partial u_r}{\partial\theta}, \qquad \gamma_{z\theta} = \frac{1}{r}\frac{\partial w}{\partial\theta} + \frac{\partial u_\theta}{\partial z}$$

and

$$\gamma_{rz} = \frac{\partial u_r}{\partial z} + \frac{\partial w}{\partial r}$$

PROBLEMS

7-1 Plot the maximum shear stress versus distance (in multiples of the lattice parameter a_o) for a straight-edge dislocation in a copper crystal. Property values are: $G = 50$ GPa; $b = 0.256$ nm; $v = \frac{1}{3}$. Show that at a distance of 1 micron, the stress has dropped to about 3 MPa.

7-2 Slip in chromium (bcc with atomic radius of 1.25 Å and $G = 11.2 \times 10^{11}$ dynes/cm^2) causes a step at the crystal surface of 1 micron in height. How many parallel dislocations moving on the same slip plane must emerge from the crystal to cause this step?

7-3 A polycrystalline specimen of nickel has an initial dislocation density of 10^{10}/m^2. Consider all dislocations to be parallel, of the edge type, and of the same sign. Also, assume the initial dislocations do not move but contribute to the generation of new dislocations. If the average grain diameter is 2.5

microns, find the dislocation density after a shear strain of 10 percent. State any further necessary assumptions.

7-4 Two edge dislocations of unlike sign are on parallel glide planes 50 nm apart; they are separated in the other direction by 150 nm. The material is iron.
(*a*) Will they tend to attract or repel?
(*b*) Determine the force they exert on each other when they are spaced as indicated.
(*c*) What is their final equilibrium position?

7-5 Two straight screw dislocations lie in a single grain of polycrystalline copper whose grain diameter is 20 microns. If they are of the same sign and lie 500 Å apart, find the direction and magnitude of the force, in dynes, of each dislocation upon the other.

7-6 An edge dislocation moving in an iron crystal is stopped along its full length by a barrier. A second dislocation, generated by the same source, moves along the same slip plane. How far apart will the dislocations be when the applied shear stress is 10^8 dynes/cm²?

7-7 Three parallel dislocations running perpendicular to the plane of the paper are located as shown below. This is a bcc metal where $a_o = 2.60$ Å, $G = 10^{11}$ dynes/cm², $\nu = 0.3$. If all dislocations are 15 microns in length, calculate the total force in dynes acting on each dislocation.

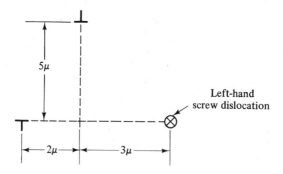

7-8 The plot of the equation describing the forces between edge dislocations on parallel planes shows a maximum and minimum for fy/C as a function of variations along the abscissa. At what values of the abscissa do these max and min points occur? See Fig. 7-19.

7-9 Two edge dislocations, A and B are of opposite sign, lie on parallel planes and are pinned so they can be considered unable to move for this problem. This crystal has properties of: $G = 10^{12}$ dynes/cm²; $b = 2.5$ Å; $\nu = 0.25$; ($\mu = 1$ micron). The sketch shows a third dislocation C lying on a plane between and parallel to the other two; note that C has the same sign as A.
(*a*) Based only upon the effects of A and B, would C move to the right or left of the position shown in the sketch, assuming no external stress was applied? You must support your answer in some definite manner.

(b) Once *C* moves as determined in step 1, where will it finally stop on its glide plane with respect to either *A* or *B*? A *reasonable* estimate, *supported by calculations*, will be adequate.

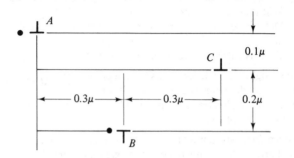

7-10 Consider a metal whose properties include: $E = 10^7$ psi (elastic modulus); $v = 0.3$ (Poisson's ratio); $r = 1.25$ Å (atomic radius). Suppose a portion of a dislocation is pinned between two sites that are 100 Å apart centrally located in a crystal whose diameter is about 10^{-4} cm. Approximately what minimum applied shear stress is required to start a source generating dislocations on the glide plane?

7-11 Fine particle strengthening can be analyzed in terms of a Frank-Read model. Aluminum alloy 7075 in the T6 condition (i.e., precipitation hardened) has a tensile yield strength of about 85 ksi and the average particle spacing is about 50 nm. Determine the average particle spacing that would be predicted as based upon an analysis using the aforementioned model.

7-12 Three edge dislocations pile up in an equilibrium position as shown in the sketch; the pile up is caused by a barrier and the applied (i.e., external) shear stress is 2×10^9 dynes/cm². If the material is iron, determine the distance indicated as *x* and *y*. Note that $G = 8(10)^{11}$ dynes/cm²; $v = 0.3$; and $b = 2.48$ Å.

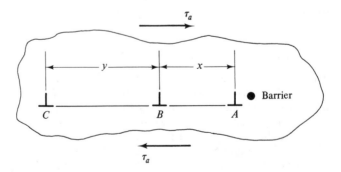

7-13 Two parallel edge dislocations of opposite sign lying parallel to the z direction in an iron crystal are subjected to σ_{yx} of 125 MPa. Calculate the distance x_o between the dislocations for which there is no net force on the dislocations, assuming A is pinned in place and B moves to the left.

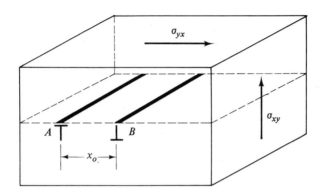

7-14 Calculate the stress needed to start a Frank-Read source in a magnesium crystal containing fine hard particles that are 0.43 nm in diameter and with a mean spacing of 74 nm.

7-15 The shear modulus of a metal is about 12×10^6 psi, has a Poisson's ratio of 0.30, and a Burgers vector of 2.6 Å. In the sketch, suppose dislocation A is pinned and B initially lies at position 1.
(a) Describe the events that occur, including the determination of any necessary *applied* τ, as B is moved to point C.
(b) Repeat part (a) above if B is to be moved from point C to point 2.
(c) What is the single largest value of τ that must be applied to move B from point 1 to point 2?

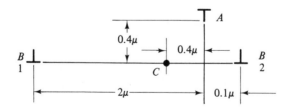

7-16 Three edge dislocations, all of the same length, lie on parallel glide planes as shown where A and C are of like sign while B and C are unlike. Assuming A and B are pinned, what will be their effect on C as to its movement (i.e., will it stay where it is, move to the right or to the left)? You must give supporting calculations to justify your answer. For this situation, $C = Gb^2/2\pi(1-\nu)$ is given as 1.2×10^{-9} N or 1.2 nN. Note that the various dimensions in the sketch are given in relative terms.

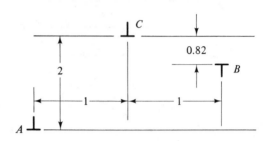

8

FRACTURE AND FRACTURE MECHANICS

8-1 INTRODUCTION

The phenomenon of fracture has been observed by all of us, and depending upon the intent involved, it may be desirable or not. Machining operations that produce certain surface geometries, the chopping of wood, and the cutting of glass to desired sizes all illustrate intentionally induced fractures. The breaking of plates during dishwashing, the accidental cracking of windshields on automobiles and the catastrophic breaking up of large ships such as oil tankers typify undesirable fractures. Regardless of how and why a fracture occurs, the end result is a creation of new surfaces and, in the extreme, at least two physical pieces from what was once a single solid. Fracture demands the existence of a crack or flaw somewhere in the solid, and stresses which induce the crack to propagate.

It is, apparently, impossible to produce a solid body, regardless of the particular material involved, so perfect in structure that it contains no flaws or microcracks of any kind.* The origination of macroflaws can arise from sources such as inclusions, whereas certain models involving the concepts of dislocation theory have been used to explain how microcracks could originate. This indicates one area of study in this complex subject and the main intent is to explain in the most basic sense why and how cracks originate. Another approach to this subject falls under the heading of Linear Elastic Fracture Mechanics (LEFM), where the stress fields around crack tips of various sizes and shapes are described. It is essentially mathematical in nature and is employed as a design tool in those situations where a traditional strength of materials approach has been found to be lacking in terms of completeness. Still another aspect of fracture is commonly called failure analysis. Here, fracture has taken place and by a careful study of the characteristics of the fracture surface, the reasons why the fracture occurred are hopefully determined. This information is then used to prevent similar fractures in the future. Note that it also plays a large role, in terms of testimony by expert witnesses, in many legal cases. From these introductory remarks, it can be seen that there are quite different interests in and approaches to the topic of fracture. It is intended in this chapter to present at least the major aspects of these topics in an introductory sense; if the reader wishes to pursue a specific subject in greater detail, more specialized reference texts must be consulted.

8-2 MODES OR TYPES OF FRACTURE

As a reasonable generalization, fracture may be categorized as brittle or ductile. In this context, the behavior is related to the strain at which fracture occurs. With brittle fracture there is little if any permanent deformation and the shape change is minimal. Solids like glass and gray cast iron fit this category when parameters such as temperature and pressure are restricted to so-called normal conditions. Because small strains are involved, the pieces could be refitted and the overall shape would be the same as that prior to fracture. Gluing a chip from a piece of tableware indicates this general concept. Ductile fractures, however, are preceded by appreciable plastic deformation before the actual separation occurs. Although the pieces might be refitted. the overall shape may be vastly different from the initial state. Most actual fractures involve both modes but are usually dominated by one. In a crystallographic sense, brittle fractures usually occur by cleavage, where the tensile stresses literally pull apart adjacent planes of atoms; ductile fractures usually occur because of shear stresses which cause atoms to slip with respect to

*Metallic whiskers are a reasonable exception.

each other. From the appearance of the fracture surface, the word "fibrous" is often used to describe ductile behavior while "granular" describes the brittle type. For the purposes of this text, the words ductile and brittle will be utilized and should be considered to describe the relative magnitudes of induced strain prior to the fracture itself.

Because of its simplicity and wide use, a standard tensile test best serves to display the fracture modes. Consider Fig. 8-1(a) which describes the general behavior of gray cast iron or glass under uniaxial tensile stresses. The test terminates with no apparent shape change of the specimen and the area under the σ-ϵ curve implies that a certain amount of strain energy was stored in the specimen up to the onset of fracture. Although with some brittle solids the fracture stress σ_f may be relatively high, the very small value of the fracture strain ϵ_f means that the energy induced prior to the fracture must be small. Because of this, brittle fractures require low energy to failure. Note that the fracture surface is essentially normal to the applied tensile stress.

Figure 8-1(b) displays the type of behavior seen with ductile materials such as lead or aluminum. In this instance a large reduction of area occurs and much larger amounts of strain energy are expended prior to the actual fracture. This is indicated in a qualitative manner by the relatively large area under the σ-ϵ curve. The cup and cone fracture of metals such as annealed, low-carbon steels is shown in Fig. 8-1(c).

One of the disturbing problems in engineering design concerns the brittle failure of materials that are generally considered as being ductile. The most startling example is the sudden, and usually catastrophic, breaking in two of large steel tankers. Evidence such as surface appearance and the absence of gross permanent deformation certainly point to a brittle fracture, yet laboratory tests of the material adjacent to the fracture surface have showed the usual ductility.

Various parameters such as temperature and strain rate can be altered so as to have a normally ductile metal fail in a brittle manner yet they alone cannot fully explain all of the situations involving such fractures.

As a reasonable general rule, it is well known that for a given metal, factors which raise the stress-strain curve tend to promote brittle fracture of a normally ductile material. Figure 8-2 illustrates the concept. Imagine that a *fracture curve* exists for a particular material and the intersection of the σ-ϵ curve with the imaginary one would cause conditions defining fracture. Under conditions indicated by curve A, fracture would occur after a large amount of energy was expended; this would correspond to a relatively high fracture strain shown as ϵ_{f1}. Testing a similar specimen at lower temperatures, higher strain rates and/or greater triaxial stresses will tend to raise the σ-ϵ curve as indicated by points B through D. It is obvious that the total energy to fracture, seen as the area under the curve, is lowered as is the

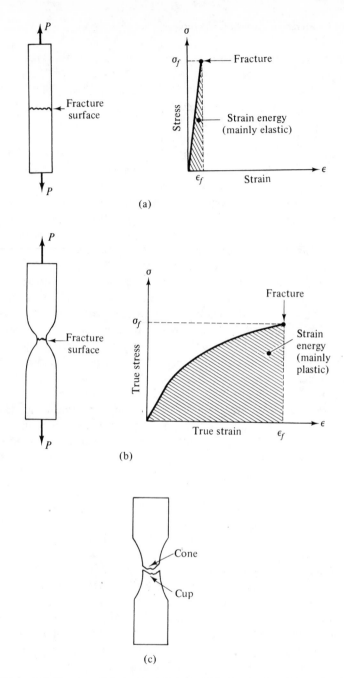

Figure 8-1 Stress-strain behavior of brittle solids (a), ductile solids (b), and a cup and cone type of fracture (c).

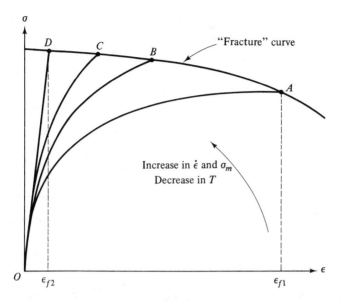

Figure 8-2 Influence of factors that tend to promote brittle fracture of a normally ductile solid.

fracture strain shown as ϵ_{f2}. This results in a more brittle behavior. Figure 8-3 is a three-dimensional plot that portrays such results more fully.

It is now appropriate to consider fracture at the atomic level.

8-3 MAXIMUM THEORETICAL COHESIVE STRENGTH OF SOLIDS

In the classical sense Fig. 8-4 describes the force reactions between two free atoms as the interatomic spacing is varied. Both repulsive and attractive effects occur and in combination produce the summation force curve. At the point defined by a_o the net force is zero and this distance is considered to be the atomic spacing at which equilibrium occurs. To decrease a_o, by inducing compression, reveals a significantly increasing repulsive force to be overcome. To increase a_o, by inducing tension, it can be seen that the force summation curve first reaches a maximum and then decreases steadily as the interatomic spacing is increased. It is useful to think of this behavior in terms of developing a theoretical cohesive strength such that if the peak force in the summation curve were actually applied, then the interatomic spacing would simply increase indefinitely. In such a context, by replacing the two atoms with adjacent rows of atoms, the application of some critical

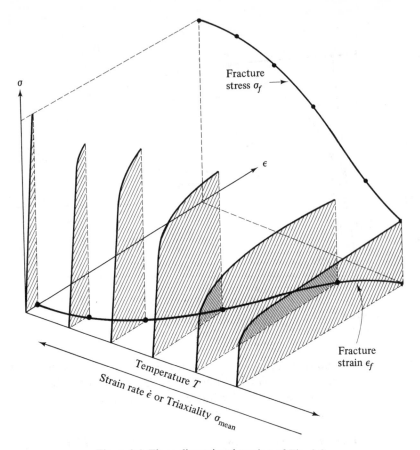

Figure 8-3 Three-dimensional version of Fig. 8-2.

force or stress would literally tear the adjacent rows apart and produce two new surfaces.

As a mathematical model, first assume that the shape of the summation curve in Fig. 8-4 is described by Fig. 8-5, where the ordinate is in terms of applied stress rather than force. An increase in the applied tensile stress, starting from the unstrained condition designated by a_o, finds repulsive effects decreasing more rapidly than attractive effects. Once point A is reached and then exceeded, the stress needed to continue increasing the atomic spacing becomes less and less. Thus, the maximum cohesive strength is interpreted as σ_m. Although several approaches have been followed to determine the magnitude of σ_m, they all produce fairly similar results. For our purposes, assume that the shape of the curve in Fig. 8-5 follows a sine function between a_o and $a_o + (\lambda/2)$ where λ is a fictitious wave length. Thus,

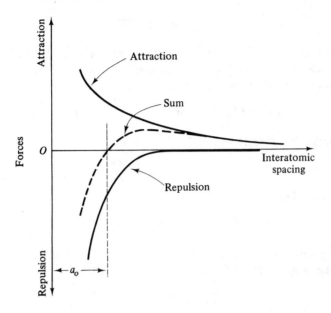

Figure 8-4 Effect of atomic spacing on force resulting between two atoms.

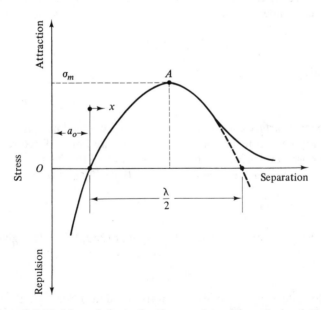

Figure 8-5 Model used for estimating maximum theoretical cohesive strength.

$$\sigma = \sigma_m \sin \frac{2\pi x}{\lambda} \tag{8-1}$$

Now for small values of displacement, shown as x, assume that such behavior is reasonably described by Hooke's law and that a_o is the initial length. Since small strains may be defined as $\Delta l/l_0$ we find:

$$\sigma = E\epsilon = E\left(\frac{x}{a_o}\right) \quad \text{or} \quad \frac{d\sigma}{dx} = \frac{E}{a_o} \quad \text{as } x \longrightarrow 0 \tag{8-2}$$

Differentiating Eq. (8-1) with respect to x gives:

$$\frac{d\sigma}{dx} = \frac{2\pi\sigma_m}{\lambda} \cos \frac{2\pi x}{\lambda} \tag{8-3}$$

which as $x \longrightarrow 0$ gives:

$$\frac{d\sigma}{dx} = \frac{2\pi\sigma_m}{\lambda} \quad \text{since the cos term} \longrightarrow 1 \tag{8-4}$$

Equating Eqs. (8-2) and (8-4) results in $(2\pi\sigma_m)/\lambda = E/a_o$. Since λ and a_o are of the same order we find:

$$\sigma_m \approx \frac{E}{2\pi} \quad \text{or approximately} \frac{E}{10}$$

Note the similarity with the derivation for the maximum theoretical shear strength in the previous chapter where $\tau_m = G/(2\pi)$ was given by Eq. (7-5).

The work required to cause fracture on a unit area basis may be determined from the area under the curve in Fig. 8-5. For all practical purposes, this is:

$$W_f = \int_0^{\lambda/2} \sigma_m \sin \frac{2\pi x}{\lambda} \, dx \tag{8-5}$$

which leads to

$$W_f = \frac{\sigma_m \lambda}{\pi} \tag{8-6}$$

Here, the term *surface energy* is introduced and denoted as γ. In essence this is the energy needed to produce a new surface and every fracture is viewed as causing *two* new surfaces in this context. If all of the fracture work is used to create two new surfaces only then with $W_f = 2\gamma$ and using Eq. (8-6):

$$\sigma_m = \frac{2\pi\gamma}{\lambda} \tag{8-7}$$

Equating Eq. (8-7) and the earlier finding that $(2\pi\sigma_m)/\lambda = E/a_o$ gives:

$$\sigma_m = \left(\frac{E\gamma}{a_o}\right)^{1/2} \tag{8-8}*$$

If property values of real solids were introduced into Eq. (8-8), the predicted theoretical value of σ_m would be on the order of $E/10$ as indicated earlier.

*Orowan [3] apparently proposed Eq. (8-8) initially.

Example 8-1.

In regard to Eq. (8-8), reasonable property values for steel would be:

$$E = 30 \times 10^6 \text{ psi}$$
$$\gamma = 1000 \text{ ergs/cm}^2$$
$$a_o = 2.5 \text{ Å}$$

Determine the theoretical maximum cohesive strength of steel.

Solution.

Conversion to SI units gives:

$$E = 30 \times 10^6 \times 6.895 \times 10^3 = 207 \times 10^9 \text{ N/m}^2 = 207 \text{ GPa}$$

$$\gamma = 1000 \times 10^{-7} \times 10^4 = 1 \frac{\text{N} \cdot \text{m}}{\text{m}^2}$$

$$a_o = 2.5 \times 10^{-10} \text{ m}$$

Using Eq. (8-8):

$$\sigma_m = \left(\frac{E\gamma}{a_o}\right)^{1/2} = \left(\frac{207 \times 10^9 \times 1}{2.5 \times 10^{-10}} \frac{\text{N}^2}{\text{m}^4}\right)^{1/2}$$

$$\sigma_m = 28.77 \frac{\text{GN}}{\text{m}^2} = 28.77 \text{ GPa}$$

Note that this is equivalent to $E/7.2$, which compares favorably with $E/10$ as mentioned earlier.

Now let us make a short historical digression. In the mid-1930's the concepts of dislocation theory were postulated to explain why real solids do not possess a theoretical maximum shear strength. Rather than shearing full planes of atoms simultaneously, the motion of dislocations demands far lower stresses to bring about the end phenomenon called slip. A number of years *earlier*, A. A. Griffith [1] proposed the first theory to explain the great discrepancy between the observed cohesive strength of solids and their theoretical value of $E/10$. He postulated that solids must contain very fine cracks or flaws no matter how much care is followed in producing these solids. At the tip of these cracks, the stress concentration effect causes the theoretical cohesive strength to be reached even though the apparent nominal stress is much lower than σ_m. When such a condition occurs, the crack begins to spread and crack surface area is increased; this demands an input of energy to account for the increase in γ. This is provided by the elastic strain energy that was stored in the body prior to the onset of crack propagation. One way of stating the Griffith criterion is that a crack will propagate when the decrease in elastic strain energy is at least equal to the energy needed to create the new surfaces associated with the crack. To determine the strain energy released as a crack propagates, requires a knowledge of the stress distribution around the crack tip. Griffith used a solution developed earlier by Inglis [2] who considered the stress distribution near the end of the major axis of an elliptical hole; this was based upon elasticity theory.

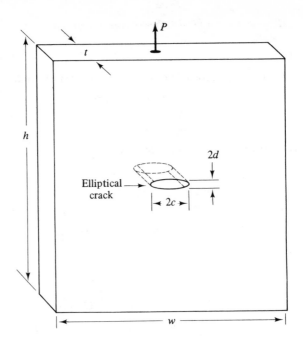

Figure 8-6 Dimensional schematic of specimen related to the Griffith analysis for plane stress.

Reference to Fig. 8-6 shows the type of specimen used to investigate the above postulate. If the thickness t is small compared to the dimensions w and h, the applied load, P, causes a condition of *plane stress* to result. For the conditions that $2c \gg 2d$ and $w \gg 2c$ and based upon the stress analysis by Inglis, Griffith determined that the elastic strain energy released is:

$$\Lambda = \frac{\pi c^2 \sigma^2 t}{E} \tag{8-9}$$

Due to the presence of the crack, the surface energy is:

$$U_s = -(2)(2c\gamma)t = -4c\gamma t \tag{8-10}$$

since there are two surfaces associated with the crack. Here any strain energy released is viewed as work being done by the system and is taken as positive in the usual thermodynamic sense; an increase in surface energy is work done on the system and is so defined by the negative term. Any change in total energy of the system that comes about due to crack propagation is then described by.

$$\Delta U = \Lambda + U_s \tag{8-11}$$

In terms of the Griffith criterion, a crack will start to propagate when U_s balances Λ or when the rate of change of U with respect to c is zero. Using this concept along with Eqs. (8-9) through (8-11),

$$\frac{d}{dc}(\Delta U) = 0 = \frac{d}{dc}\left[\frac{\pi c^2 \sigma^2 t}{E} - 4c\gamma t\right] \tag{8-12}$$

which gives:

$$\sigma = \left(\frac{2E\gamma}{\pi c}\right)^{1/2} \tag{8-13}$$

Here, σ is the stress needed to *start* a crack of starting length $2c$ to propagate, while E and γ are the elastic modulus and surface energy of the particular solid in question. Note that if this critical condition is met and σ is kept constant, the crack should continue to increase in a relatively uncontrolled manner. Later it will be shown that for the geometries involved, Griffith type cracks will always be unstable. It is of interest to note that since $(2/\pi)^{1/2} \approx 1.0$, there is a great similarity between Eqs. (8-13) and (8-8). In essence, if cracks had initial lengths on the order of the lattice parameter, a_o, maximum theoretical cohesive strengths could be approached. In real-life situations, $c \gg a_o$ and, because of this, Griffith introduced the basic reason why such theoretical strengths are not attained in practice.

Another form of Eq. (8-13) that often appears in the literature relates to the case where plane strain conditions exist. This would be for the geometry where the thickness t is large and the result is given by:

$$\sigma = \left[\frac{2E\gamma}{(1 - v^2)\pi c}\right]^{1/2} \tag{8-14}$$

where v is Poisson's ratio. For many metals, where $v \approx 0.3$, it can be seen that the difference in predictions using Eqs. (8-13) or (8-14) is quite minimal.

Example 8-2.

The sketch, not drawn to scale, shows a sheet of glass of designated dimensions containing a small crack or slit as indicated. A mass of 500 kg induces a tensile stress in the sheet and it is assumed here that any effects of the mass of the glass is included in the 500 kg value. Consider the following property values, $E = 60$ GPa, $\gamma = 0.5$ J/m², $v = 0.25$, $\sigma_f = 170$ MPa (fracture strength of sound glass). What is the minimum length of the slit that would lead to fracture under these conditions?

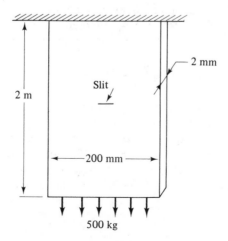

Solution.

If this is considered as a condition of plane stress, then with Eq. (8-13):

$$c = \frac{2E\gamma}{\pi\sigma^2}$$

and the various values in SI units are, $E = 60 \times 10^9$ N/m^2 and $\gamma = 0.5$ N·m/m^2.

$$\sigma = \frac{500(9.8)}{200(2)(10^{-6})} \frac{N}{m^2} = 1.23 \times 10^7 \frac{N}{m^2}$$

$$c = \frac{2(60) \times 10^9(0.5)}{\pi(1.23)^2 \times 10^{14}} = 1.26 \times 10^{-4} \text{ m} = 0.126 \text{ mm}$$

So, the crack length $2c$ is 0.252 mm.

If plane strain conditions are considered then with Eq. (8-14)

$$c = \frac{0.126}{(1 - \nu^2)} = 0.134 \text{ mm}$$

Note that the plane stress calculation is more restrictive (i.e., allows a smaller starting crack) but is probably more correct in a physical sense.

Students often assume that they must make use of all information provided in problems and use the fracture strength in this situation in place of the applied stress. As engineers, it will be their responsibility to decide what information is both useful and necessary when confronted with a problem. Thus, providing extraneous information in home assignments and on examinations will, hopefully, begin to develop the ability to select pertinent data and to exclude that which is not essential in a particular situation. In this way, simple formula plugging can be overcome.

The proper use of Eqs. (8-13) or (8-14) is restricted to solids whose fracture requires only enough input energy to overcome the surface energy, γ. Few real materials satisfy this constraint; in fact, even with metals such as gray cast iron, which is certainly brittle in a general sense, it has been shown that some *plastic* deformation accompanies the fracture process. To account for this, both Orowan [3] and Irwin [4] proposed a modification of the Griffith equation. The modified Griffith equation is expressed as:

$$\sigma = \left(\frac{2E(\gamma + P)}{\pi c}\right)^{1/2} \approx \left(\frac{EP}{c}\right)^{1/2} \tag{8-15}$$

where P is the plastic work needed to be overcome if the crack is to increase. Although P is difficult to determine with great accuracy, reasonable estimates indicate that $P \gg \gamma$ so that an approximate form which includes this modification is often written as shown in Eq. (8-15). Both γ and P may be viewed as quantities that sop up energy as fracture proceeds and to determine their individually distinct quantities poses considerable difficulties.

8-4 STRAIN ENERGY RELEASE RATE

As an alternative approach, the concept of the *strain energy release rate* has been introduced via a parameter denoted as G.* In effect, $G = 2(\gamma + P)$, so Eq. (8-15) is written as:

$$\sigma = \left(\frac{EG}{\pi c}\right)^{1/2} \tag{8-16}$$

Perhaps the principal advantage in introducing G is that it is easier to obtain experimentally as compared to the individual determination of γ and P. Physically, G is related to the source of total energy which is needed to cause crack extension once a critical situation is reached. Its proper dimensional units are energy per crack area, that is, the energy required to increase the crack by some unit area. Although it is often called the crack extension force, this description is discouraged since it is not in reality a force. When G reaches a critical value, usually denoted as G_c, an instability condition is reached and crack propagation occurs. To determine G_c quantitatively, it is necessary to run experiments; the development that follows shows how G_c is related to measurable parameters. Consider the case for a *fixed grip* situation indicated by Fig. 8-7.

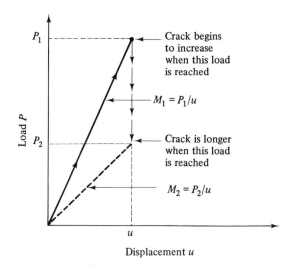

Displacement u

Figure 8-7 Load-displacement plot for crack extension at constant displacement showing change in compliance, $1/M$.

*Do not confuse this with the shear modulus.

The strain energy

$$\Lambda = \tfrac{1}{2}Pu \quad \text{and} \quad P = Mu \tag{8-17}$$

Note that since u is constant, P/M is constant, thus the decrease in the load under crack extension is accompanied by a proportional change in the compliance, $1/M$. It then follows that:

$$\frac{\partial \Lambda}{\partial A}\bigg]_u = \frac{1}{2}u\frac{\partial P}{\partial A}\bigg]_u = \frac{1}{2}\frac{P}{M}\frac{\partial P}{\partial A}\bigg]_u \tag{8-18}$$

Since $P/M = u = $ constant,

$$\frac{1}{M}\frac{\partial P}{\partial A} + P\frac{\partial(1/M)}{\partial A} = 0 \tag{8-19}$$

Combining Eqs. (8-18) and (8-19) gives:

$$\frac{\partial \Lambda}{\partial A}\bigg]_u = -\frac{1}{2}P^2\frac{\partial(1/M)}{\partial A} = G_c \quad \text{at instability} \tag{8-20}$$

If a similar analysis were performed with a constant load condition as in Fig. 8-8, the same result would be found except that a positive sign would appear in Eq. (8-20). This implies that the same energy release rate is needed

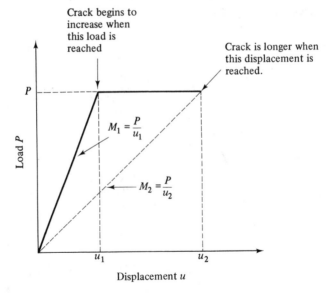

Figure 8-8 Same as Fig. 8-7 except crack extension occurs at constant load.

to initiate crack propagation regardless of the manner of loading which causes crack extension. Thus, the *critical* strain energy release rate is defined by Eq. (8-20) as:

$$G_c = \frac{1}{2} P_c^2 \frac{\partial(1/M)}{\partial A} \tag{8-21}$$

P_c being the critical load and $1/M$ the compliance.

To determine G_c experimentally, it is often necessary to determine the influence of crack area on the compliance. Suppose that into a series of identical specimens, individual cracks of different starting lengths are introduced (i.e., a different crack length per specimen). Each specimen will then exhibit a particular value of $1/M$ per crack length. This provides a plot of $1/M$ versus crack length. By then subjecting a single specimen with known starting crack length to an ever-increasing load, the value of P_c to initiate crack propagation may be found. By determining the slope of the compliance–crack length curve at that crack length for which P_c was determined, these particular values may be introduced into Eq. (8-21) to determine G_c. This procedure is often called the *calibration bar technique*.

Example 8-3.

A number of identical specimens of a given metal differ only in that the starting central crack is cut to different lengths. The specimens are subjected to tensile loads and the compliance versus starting crack length is determined for each. These results are then plotted in the form of a curve of $1/M$ versus crack length. From these results, the *slope* of this curve at a crack length of 0.100 in. is found to be 10^{-4} lbf^{-1}. A subsequent test, using a specimen having a starting crack of 0.100 in., shows that the load required to start crack propagation is 300 lbf. Determine G_c if all specimens had a thickness of 0.375 in.

Solution.

From Eq. (8-21):

$$G_c = \frac{1}{2} P_c^2 \frac{\partial(1/M)}{\partial A} \qquad \text{where } \partial A = t \, \partial c$$

since t is constant (*Note:* Here c is the *full* crack length). Now $P_c = 300$ lbf, $t = 0.375$ in., and $[\partial(1/M)/(\partial c)] = 10^{-4}$ lbf^{-1} so:

$$G_c = \frac{1}{2}(300)^2(10^{-4})\left(\frac{8}{3}\right)\frac{\text{lbf}}{\text{in.}}$$

$$G_c = 12\,\frac{\text{lbf}}{\text{in.}} \qquad \text{or, more properly, } 12\,\frac{\text{in.-lbf}}{\text{in}^2}$$

Note: Problem 8-11 poses a more complete example.

8-5 DESIGN CONSIDERATIONS

Based upon a Strength of Materials (SM) approach to the design of structures, it is usual to introduce a factor of safety whereby the expected allowable stresses to be resisted are taken as some percentage of the yield or tensile strength of the material to be used. Although this safety factor is often ill-defined in a fundamental sense, the structure is, hopefully, overdesigned and should, therefore, behave in a satisfactory manner. If it were possible to introduce G_c directly into the SM approach, the design against fracture would be entirely different from what it is. Unfortunately, this is not the case and the basic reason lies in the complete absence of crack influence in the equations available in this method of design. Thus, although G_c may be viewed as a material property in the same way one considers yield strength, by itself it cannot be used in a direct, quantitative manner. As will be seen shortly, its physical meaning does have direct bearing in designing against brittle fracture.

To this point, discussion and mathematical derivations have constantly emphasized crack size and propagation. Why, it might be properly asked, has the SM approach been used successfully for many years when the influence of cracks has been ignored? The answer is that in a fortuitous sense, many structures and other components have been made of materials having the inherent ability to plastically deform around the tip of cracks and thereby negate further unstable propagation. A comparable question would be to ask why have numerous other structures, made of apparently ductile materials, failed in a brittle manner. It is in the latter situations that one must seek a superior alternative to SM design concepts and that is what the subject of fracture mechanics is really all about.

8-6 LINEAR ELASTIC FRACTURE MECHANICS

In real structures, flaws or cracks can originate for a variety of reasons. Defects due to welding, the effects of stress corrosion or even the presence of microcracks explained by dislocation theory indicate potential sources. The existence of a crack by itself is not of true concern; it is only when service conditions cause the crack to propagate that concern may turn to dismay. Starting with the premise that structures will contain flaws or cracks, it becomes essential to consider their size, shape, etc. in initial design considerations; this introduces the key difference between the concepts of fracture mechanics and strength of materials.

It is now generally accepted that the primary parameters involved in brittle fracture design are the size, shape, and location of the worst crack, the magnitude of applied tensile stresses, and a material property called fracture toughness.* Other variables such as temperature and strain rate, although of decided influence, may be viewed as secondary parameters in exactly the same vein in which they influence yield strength or other material properties.

---

*Different definitions and symbols are used for this "property" so care must be exercised.

As an aside, but an important one, *qualitative* comparisons of materials involving the energy to cause fracture have been made for years using the Charpy V Notch test (CVN). Certainly much useful information related to transition from ductile to brittle behavior as a function of temperature has been obtained from this test. However, such numerical test values have not yet found use in a direct analytical manner that would lead to *quantitative* predictions. This is the primary limitation of such tests. Figure 8-9 shows a typical Charpy plot.

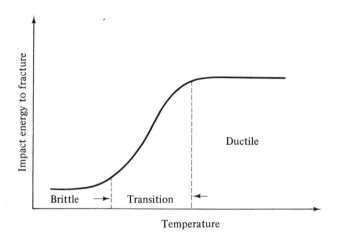

Figure 8-9 General shape of Charpy plot.

Linear Elastic Fracture Mechanics (LEFM) introduces, by analytical means, the equations which describe the magnitude and distribution of stresses around the tip of a crack as a function of the applied stresses, crack size and shape, and a parameter called the *stress intensity factor*. The latter, denoted as K, is considered as a scale factor that accounts for the presence of a crack as it affects the stresses in the vicinity of the crack tip.* The importance of K is that it describes the stress field, not simply the largest single stress, and when it reaches a critical value, K_c, crack extension is imminent.

Traditionally, three possible loading modes have been considered in regard to crack opening. They are called Modes I, II and III as shown in Fig. 8-10 and the stress intensity factors for these modes are denoted as K_I, K_{II}, and K_{III}. Because the Mode I opening is encountered far more often than the other two and the majority of research studies have been devoted to this type of loading, no further detailed discussion will be devoted to Modes II or III. In fact, the bulk of the remainder of this section will be concerned with a Mode I loading under *plane strain* conditions, since at the present time it is

---

*This should not be confused with the stress *concentration* factor used in macroscopic design.

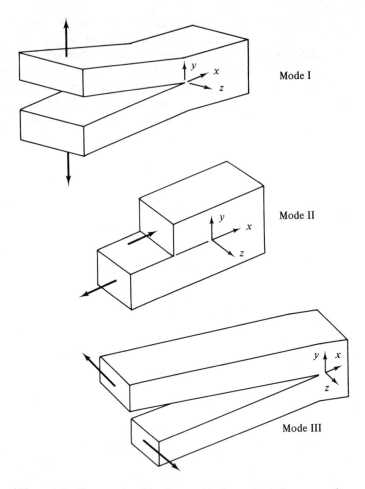

Figure 8-10 Three modes of loading used in linear elastic fracture mechanics analyses.

the only situation for which standardized test procedures have been developed and accepted [5]. The critical parameter of concern is written as K_{Ic} and will receive our major attention.*

The equations that describe the stress field for Mode I loading come from a method attributed to Westergaard [6] and are as follows:

$$\sigma_y = \frac{K_I}{(2\pi r)^{1/2}} \cos \frac{\theta}{2} \left[1 + \sin \frac{\theta}{2} \sin \frac{3\theta}{2} \right]$$

*K_{Ic} is the plane *strain* critical stress intensity factor for static loading. Similar factors for plane stress conditions, dynamic loading and for modes II and III have also been developed [7].

$$\sigma_x = \frac{K_\mathrm{I}}{(2\pi r)^{1/2}} \cos\frac{\theta}{2}\left[1 - \sin\frac{\theta}{2}\sin\frac{3\theta}{2}\right]$$

$$\tau_{xy} = \frac{K_\mathrm{I}}{(2\pi r)^{1/2}} \sin\frac{\theta}{2}\cos\frac{\theta}{2}\cos\frac{3\theta}{2}$$ (8-22)*

$$\sigma_z = \nu(\sigma_x + \sigma_y), \qquad \tau_{xz} = \tau_{yz} = 0$$

where Fig. 8-11 defines the notations of the coordinate system used.

As mentioned earlier, the case of plane strain will receive major attention and Eq. (8-22) expresses that physical situation where higher order terms in r have been omitted. As r approaches the edge of the crack, these equations provide a good prediction of the stresses in that region. Of course it is obvious that these predictions indicate the possibility of extremely high stresses in the vicinity of the crack tip (i.e., they approach ∞ as $r \rightarrow 0$) and could not be expected in a real material. The limit to any actual stress magnitude would be the yield strength of the solid. That is, regardless of what values would be *predicted* by Eq. (8-22), the maximum value would *cut off* when yielding occurred. Figure 8-12 illustrates this physical situation for σ_y at $\theta = 0$. As may be seen in Eq. (8-22) K_I is related to stress and the square root of some characteristic length so in terms of dimensional correctness, it must have units of stress times (length)$^{1/2}$ (e.g., psi$\cdot$(in.)$^{1/2}$ or Pa$\cdot$m$^{1/2}$). It is now well

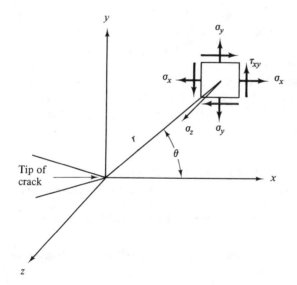

Figure 8-11 Stress state in the vicinity of a crack tip showing coordinates of concern.

*In earlier notation, still found in other sources, $\mathcal{K}$ is employed. Probably because of simplicity, K has found increased usage; they are related by $K = \mathcal{K}(\pi)^{1/2}$. In this text, K will be used.

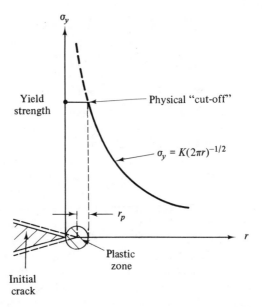

Figure 8-12 Formation of plastic zone at a crack tip when yielding occurs.

established that K_I is related to the applied stress, σ_a, and the square root of some function of the crack length, a; thus in a general form:

$$K_I = C\sigma(a)^{1/2} = f(\sigma, a) \qquad (8\text{-}23)^*$$

The functional coefficient depends upon the geometry of the body subjected to stress σ and the crack itself. Many excellent analyses have provided the solutions for K_I and can be found in several of the references [8–11]. In Fig. 8-13 some of the more widely used solutions are presented for illustrative purposes [11].

Example 8-4.

Published values indicate that for a 4340 steel, the values of K_{Ic} and the corresponding yield stress are about 100 MPa·(m)$^{1/2}$ and 860 MPa respectively. Estimate the radius of the plastic zone that would be expected to develop ahead of the crack tip when this metal is subjected to Mode I loading.

Solution.

From Fig. 8-12 the value of σ_y that is reached upon yielding is taken as the yield stress of the material.

$$r_p = \frac{1}{2\pi}\left(\frac{K_{Ic}}{Y}\right)^2$$

$$r_p = \frac{1}{2\pi}\left(\frac{100}{860}\right)^2 \text{ m} = 2.15 \times 10^{-3} \text{ m} = 2.15 \text{ mm}$$

---

*Although c described crack length in the earlier literature, a is now used. This latter symbol will be employed for the rest of this chapter.

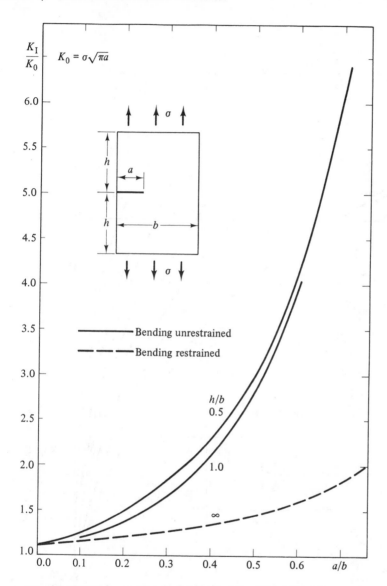

Figure 8-13a K_1 for an edge crack in rectangular sheet subjected to a uniform uniaxial tensile stress (From Rooke and Cartwright [11] with the permission of the Controller of Her Britannic Majesty's Stationery Office.).

Note: Strictly, this is for plane stress conditions. The value for plane strain conditions is about one-third of the above value. Additionally, the plastic zone size varies with angular orientation, θ, and the overall shape becomes noncircular. See Rolfe and Barsom [7] or Hertzberg [12] for a more detailed discussion.

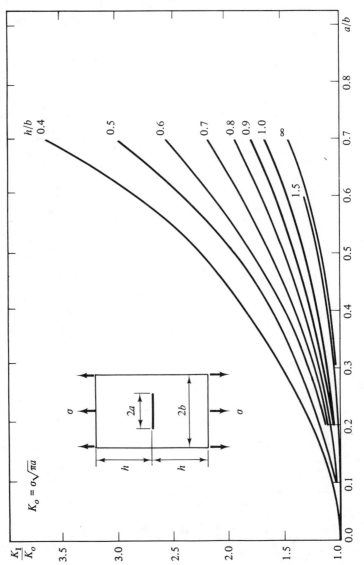

Figure 8-13b K_1 for a central crack in a rectangular sheet subjected to a uniform uniaxial tensile stress.

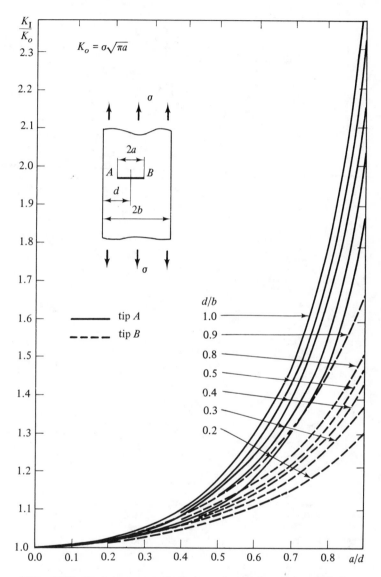

Figure 8-13c K_I for an eccentrically located crack in a finite width sheet subjected to a uniform uniaxial tensile stress.

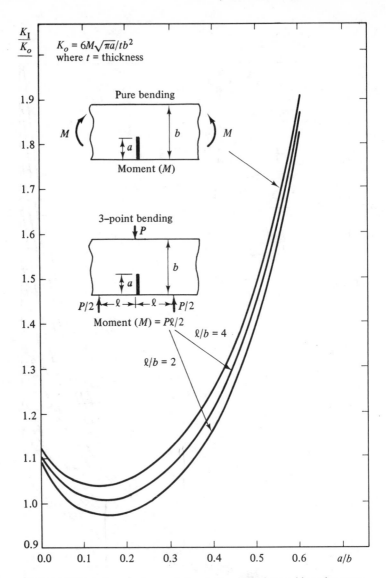

Figure 8-13d K_1 for an edge crack in a finite width sheet subjected to pure bending or three-point bending.

The design concept that enters may now be described. Once a material is selected for some application, its properties are fixed and, in essence, some critical value of the stress intensity factor, K_{Ic}, is established.* Depending

---

*As with other mechanical properties, this is determined by experiments. Although K_{Ic} is now called fracture toughness in current practice, another definition is used in Secs. 8-8 through 8-11.

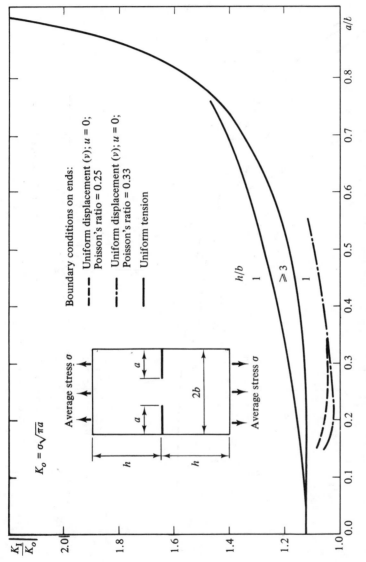

Figure 8-13e K_1 for two edge cracks in a rectangular sheet subjected to a uniform uniaxial tensile stress or a uniform normal displacement.

217

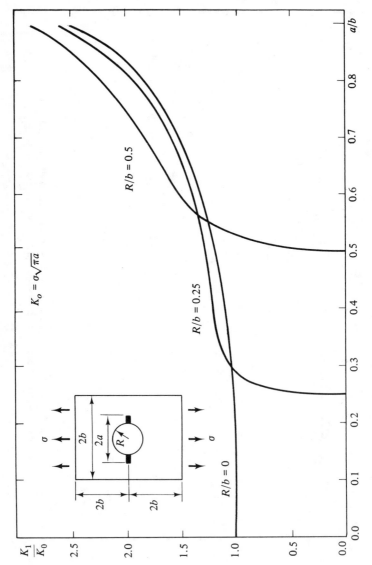

Figure 8-13f K_1 for two cracks at a circular hole in a rectangular sheet subjected to a uniform uniaxial tensile stress.

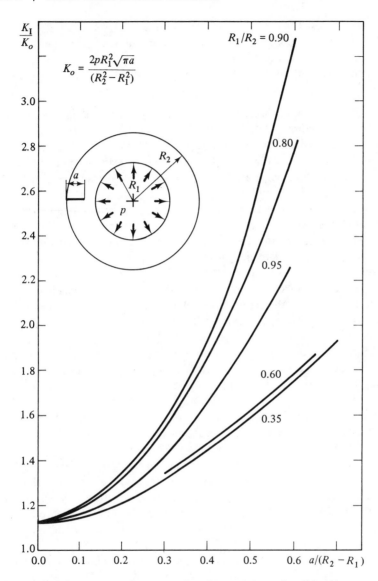

Figure 8-13g K_1 for an external radial edge crack in a tube subjected to a uniform internal pressure.

upon the specimen geometry, the functional coefficient C in Eq. (8-23) is found from the various *analyses* that led to the results typified by Fig. 8-13. Now the designer must specify the worst flaw size that can be tolerated under expected loads or σ_a, or if the worst crack size has been detected by inspection, the largest tolerable stresses can be specified. In essence for a given material-geometry combination, many combinations of stress and crack size would cause failure while many other combinations would not. Take as a simple illustration the case depicted by Fig. 8-13(a). (Ignore K_o for now.)

$$K_I = \sigma_a \cdot (a)^{1/2} \cdot (\pi)^{1/2} \tag{8-24}$$

where K_{Ic} is a property fixed by material selection; σ_a is the allowable design stress; a is the allowable crack size determined by inspection; $(\pi)^{1/2}$ is the correct functional factor determined by LEFM analysis. Thus, the designer has numerous choices that will prevent crack propagation and brittle fracture. If the structure shape and worst-known flaw size are beyond his control, a trade-off between σ_a and K_{Ic} can be made. If stresses and flaw size cannot be altered, then the selection of a material possessing an adequate K_{Ic} must be made. By adjusting the variables in a sensible manner, this design approach considers aspects not involved in a traditional SM approach and indicates why the use of LEFM techniques are superior in many situations. There is a useful analogy to remember. Where SM relates the load to σ_a to the yield strength, fracture mechanics relates the load to K_I to K_{Ic}, so as many stresses will not cause yielding, many values of K_I will not cause fracture. It is only when K_I reaches the critical value of the material property K_{Ic} that the crack will propagate.

Before the reader assumes that no further problems exist, several cautions must be added. Although numerous values of K_{Ic} may be found in the literature, there are many materials for which this property has not been measured. That is why standard experimental methods have been developed [5], although in terms of the thickness of laboratory specimens, the conditions of plane stress may be experienced.* This leads to higher values of K_c as depicted in Fig. 8-14. If such values were employed in the design of large structures where plane strain conditions exist, the actual fracture toughness, K_{Ic}, could be much lower than the laboratory value with possible catastrophic consequences. It is because K_{Ic} provides a lower and, therefore, *conservative* value that major emphasis has been geared to its measurement. Figure 8-15 illustrates the difference in the fracture surface for plane stress and plane strain behavior.

With this introduction, the reader should now be able to understand the detailed discussions on various correction factors to K_{Ic}, crack tip plastic zone size, correct dimensions of specimens used to measure K_{Ic}, etc. As a final point, there are current techniques being used to determine fracture toughness for materials that invalidate the use of LEFM. These fall in the realm of elastic-plastic behavior where the extent of plastic deformation exceeds that for which LEFM analysis can tolerate. Of the current methods, the *R*-curve Analysis, Crack-Opening Displacement (COD), and *J*-integral

---

*K_c is often used to denote the plane stress critical stress intensity factor. See Rolfe and Barsom [7] for a discussion.

are receiving most attention. To discuss these techniques is beyond the intent of this text but the reader should now be able to pursue these points by checking other texts [7, 12] of a more detailed nature.

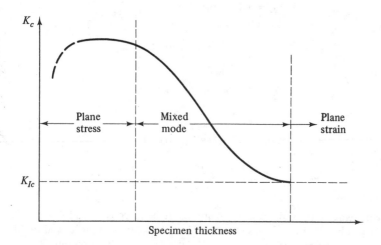

Figure 8-14 Illustration of the effect of specimen thickness upon the critical stress intensity factor.

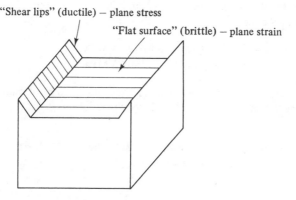

Figure 8-15 General fracture surface configuration showing the influence of plane stress and plane strain.

Example 8-5.

A cylindrical pin is made of a metal whose value of K_{Ic} is 30 MPa·m$^{1/2}$. The maximum expected service stress, including a factor of safety, is 100 ksi. Determine the maximum size surface crack that this pin can tolerate in order to prevent crack propagation. Assume the pin diameter and length are of the same size while both are much larger than the crack length.

Solution.

From Fig. 8-13, sketch (a) is most appropriate for this situation, and since both h and b are $\gg a$, $K_I/K_o = 1.1$ so

$$K_{Ic} = 1.10\sigma(\pi a)^{1/2} \text{ or } a = \frac{K_{Ic}^2}{(1.10)^2\pi\sigma^2}$$

where

$$K_{Ic}^2 = 900 \times 10^{12} \frac{N^2\text{-m}}{m^4}$$

and

$$\sigma = 100{,}000 \text{ psi} = 6.895 \times 10^8 \frac{N}{m^2}$$

Therefore,

$$a = \frac{900 \times 10^{12}}{1.21\pi(47.54) \times 10^{16}}$$

or

$$a = 0.50 \times 10^{-3} \text{ m} = 0.50 \text{ mm}$$

Example 8-6.

Two metals of the same density, A and B, are being considered for the design of a large plate that is 6 ft wide by 10 ft long. A maximum tensile load of 4.5 MN will act perpendicular to the 6-ft dimension and consideration must be given to the possibility of pre-existing cracks. For this problem, through thickness cracks at the center of the plate and perpendicular to the applied load are of major concern. Because of inspection limitations, the smallest crack that can be detected is on the order of 0.100 in. The plane strain critical stress intensity factor and yield strength of the metals are:

A. $K_{Ic} = 115 \text{ MPa·(m)}^{1/2}$, $Y = 910 \text{ MPa}$
B. $K_{Ic} = 55 \text{ MPa·(m)}^{1/2}$, $Y = 1035 \text{ MPa}$

Which material would you recommend if the plate is to be of minimum weight? In essence, a minimum thickness, t, is required.

Solution.

Two competing design considerations must be checked. The first involves yielding in the traditional design sense while the other involves the concepts of fracture mechanics.

Yielding.

A. $Y = \text{load/area} = 4.5 \text{ MN}/(t)[(6)(3.048 \times 10^{-1})] \text{ m}^2$,   so
     $t = [(4.5)(10)]/[(910)(6)(3.048)] \text{ m} = 2.7 \times 10^{-3} \text{ m} = 2.7 \text{ mm}$
B. $t = 2.7(910/1035) = 2.38 \text{ mm}$.

It should be obvious that the metal with the higher yield strength requires a smaller thickness.

Crack Growth.   From Fig. 8-13(b),

$$\frac{2a}{2b} = \frac{0.1}{72} = 0.0014 \quad \text{and} \quad \frac{2h}{2b} = \frac{10}{6} = 1.67$$

so

$$\frac{K_I}{K_o} = 1.0 \quad \text{or} \quad K_I = \sigma\sqrt{\pi a} \quad \text{where } a = 0.05(2.54)(10^{-2})\text{m}.$$

A. $\qquad \sigma = \frac{K_{Ic}}{(\pi a)^{1/2}} = \frac{115}{[(\pi)(1.27 \times 10^{-3})]^{1/2}} \text{ MPa} = 1.82 \text{ GPa}$

Since the yield stress (910 MPa) is much lower than the maximum stress allowed prior to crack growth (1.82 GPa), yielding would govern the design if metal A is used. This can also be seen by calculating the value of t based upon crack growth considerations as follows:

$$t = \frac{4.5 \text{ MN}}{(1.82 \text{ GPa})(1.83 \text{ m})} = 1.35 \text{ mm}$$

as compared with the value of 2.7 mm based upon yielding.

B. $\qquad \sigma = 1.82\left(\frac{55}{115}\right) = 870 \text{ MPa}$

which is lower than the yield stress (1035) MPa so crack growth governs the design for this metal.

$$t = 1.35 \ (1.82/0.87) = 2.82 \text{ mm}$$

as compared with the value of 2.38 mm based upon yielding.

Thus the minimum value of t for metal A that satisfies both design criteria is 2.7 mm whereas its counterpart for metal B is 2.82 mm. In this situation, metal A which has a lower Y and higher K_{Ic} is preferred on the basis of minimum weight. Two points are worth noting:

1. If only a traditional strength of materials approach were used, metal B would have been chosen and failure would occur because of uncontrolled crack growth.
2. The largest starting crack size will influence the choice of A or B since the larger the crack, the more influence the relative value of K_{Ic} has on the final decision. In this example, if a starting crack size were 0.050 in., it would be found that yielding governs and metal B would be the proper selection. Based upon yielding, there is *one* value of t that can be computed; the value of t determined from K_{Ic} values will be a function of $(a)^{1/2}$. This illustrates a key difference between these design philosophies.

8-7 TIE-IN BETWEEN STRAIN ENERGY RELEASE RATE AND THE STRESS INTENSITY FACTOR

In Sec. 8-4, reference to the strain energy release rate G was made and a brief explanation given regarding its measurement. One might wonder, in view of the section on LEFM, why any mention was made of G. In fact there is a tie-in between the two as will now be shown. By comparing Eq. (8-16) with the appropriate expression for the stress intensity factor:

$$\sigma(\pi a)^{1/2} = (EG_c)^{1/2} \quad \text{and} \quad K_c = \sigma(\pi a)^{1/2} \tag{8-25}$$

Therefore,

$$K_c = (EG_c)^{1/2} \quad \text{for plane stress} \tag{8-26}$$

In the case of plane strain,

$$K_{Ic} = \left(\frac{EG_c}{1 - v^2}\right)^{1/2} \tag{8-27}$$

Thus, if one could measure G_c for Mode I loading, the appropriate value of K_{Ic} could be computed. In current practice this is not done as the equations for standard test specimens have been developed so that K_{Ic} is computed directly. Yet to reach K_{Ic} at crack *initiation*, it could be argued that G_c has been reached concurrently. There is also a caution to be made in interpreting Eqs. (8-26) and (8-27). Due to the effect of Poisson's ratio, it would seem that plane strain values of K_c would be greater than those for plane stress. As shown in Fig. 8-14, the reverse is true. This apparent anomaly can best be explained by pointing out that the value of G_c in Eqs. (8-26) and (8-27) would not be the same for thin (plane stress) and thick (plane strain) specimens of the same material. The plane stress values would be enough higher to more than offset the $(1 - v^2)$ term in Eq. (8-27).

Example 8-7.

If the tests conducted in Example 8-3 were done under plane strain conditions and the material displayed an elastic modulus of 15×10^6 psi and Poisson's ratio of 0.35, determine the probable value of K_{Ic}.

Solution.

Using Eq. (8-27):

$$K_{Ic} = \left(\frac{EG_c}{1 - v^2}\right)^{1/2} = \left(\frac{15 \times 10^6 \times 12}{0.878}\right)^{1/2} \text{psi} \cdot (\text{in.})^{1/2}$$

$$K_{Ic} = 1.43 \times 10^4 \text{ psi} \cdot (\text{in.})^{1/2} = 15.8 \text{ MPa} \cdot (\text{m})^{1/2}$$

8-8   FRACTURE TOUGHNESS—THE GURNEY APPROACH

Although it has received very little notice and attention in the U. S. literature, the approach to fracture toughness proposed by Gurney and his coworkers [13, 14] merits mention for several reasons. First, it is more appealing to engineering students being introduced to this subject because the physical explanations involved are simpler and more direct than is the stress intensity factor concept. Second, quite simple experiments can be conducted to show how Gurney's fracture toughness is measured; in essence, this is equivalent to G_c and leads to a mathematical expression equivalent to Eq. (8-21). In fact, if mathematical expressions for the rate of change of compliance with respect to crack area (i.e., $\partial(1/M)/(\partial A)$ had been computed for various structure geometries and crack conditions *rather than* expressions for stress intensity factors, an alternative approach to fracture mechanics would exist. This is implied indirectly by the relations given as Eqs. (8-25) through (8-27); it is *not suggested* that it would be sensible to start from scratch and develop such expressions since the approach via K_{1c} is so well established. It will be shown, however, that in cases where $\partial(1/M)/(\partial A)$ can be computed with reasonable accuracy, the Gurney method provides a useful alternative.

Figure 8-16 shows a solid containing a crack of initial length a_i. As a force P is applied at the boundary any displacement is indicated as u. Assuming the body has constant thickness, t, the original crack area, A, is simply $a_i t$. In this approach the crack area is viewed as *one-sided*, unlike the idea where γ was first introduced.

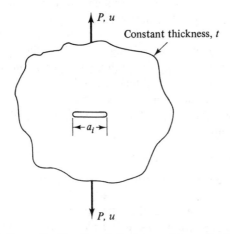

Figure 8-16 Generalized loading of a body containing an initial crack.

As P is increased, the body may be viewed as a spring whose *stiffness* is P/u or whose *compliance* is u/P either of which are dependent upon the initial value of crack length a_i. Figure 8-17 indicates how the value of stiffness would vary with initial crack length or area. If no irreversibilities occur then the loading–unloading behavior would follow an individual line as shown. Note that none of the lines is intended to describe the elastic modulus of the solid unless a_i is on the order of the flaw size existing in real solids; here it is assumed that a_i is too large to permit a proper value of the modulus to be attained although, for simplicity, the *P-u* behavior follows a straight line. At any instant, the strain energy stored in the body is directly equivalent to the area under the loading line, that is, $\Lambda = \frac{1}{2}P_1 u_1$. Note that nonlinear behavior could be handled in a similar manner as long as reversible behavior would take place upon full unloading.

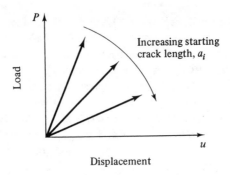

Displacement

Figure 8-17 Influence of the size of the starting crack length upon compliance or stiffness.

At some point, a critical load is reached and the crack begins to extend; thus the initial crack area, A, is increased by some amount ΔA which is equal to $(t\,\Delta a)$, t being constant. The energy or work required to produce a unit change in crack area is the fracture toughness, R^* of the material; this energy is supplied by the strain energy stored in the body. By considering the behavior of P, u, and a after the critical load is reached, the magnitude of the fracture toughness, R, may be found. A few specialized cases will be considered and in each, it will be assumed that crack propagation proceeds under *quasi-static* conditions.

CASE I: CONSTANT DISPLACEMENT

Once the crack begins to lengthen it proceeds under a falling load P but constant displacement u. This is generally called a fixed-grip type of test. Figure 8-18(a) shows the three-dimensional plot of these events while Fig. 8-18(b) shows the projection on the P-u plane. As the load drops from P_1 to P_2 under constant displacement u_1, the crack length increases by Δa (simply the new crack length a_1 minus the original length a_i).

From Fig. 8-18(b) area OAC represents the strain energy stored prior to the increase in crack length while OBC is the strain energy that remains in the body after the crack has increased to length a_1. Because u remains constant, no external work is done at the boundaries during crack extension. Thus, the change in strain energy has been consumed to overcome the fracture toughness during the increase ΔA.

Thus the sector area, OAB, is directly associated with fracture toughness as:

$$\Delta\Lambda = OAB = R\,\Delta A = R(t\,\Delta a) \qquad (8\text{-}28)$$

Note that under these conditions, the energy required to overcome R causes a *decrease* in the strain energy.

*Gurney's R is *not* the same as K_{Ic}; rather, it is analogous to G_c.

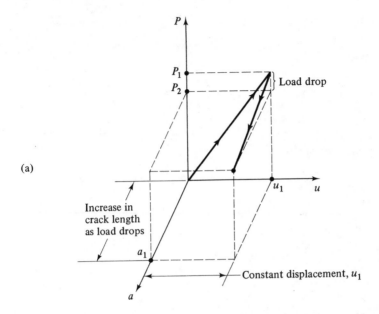

(a)

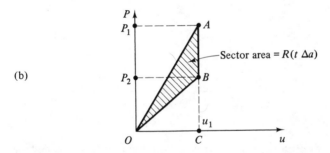

(b)

Figure 8-18 Three-dimensional plot of fracture toughness results with propagation occurring at fixed displacement (a) and projection of results onto the *P-u* plane for determining Gurney's [13] fracture toughness, *R* (b).

CASE II: CONSTANT LOAD

In this instance, the onset of crack propagation takes place under a constant load but an increasing external displacement. Figures 8-19(a) and 8-19(b) depict this situation. Because external work is not zero as in Case I, it is helpful to express this condition as:

$$\text{External work} = \Delta\Lambda + R\,\Delta A \tag{8-29}$$

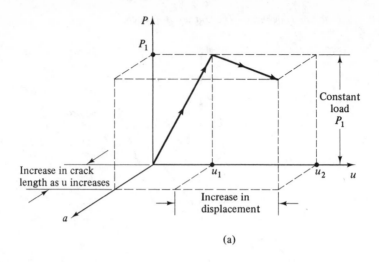

(a)

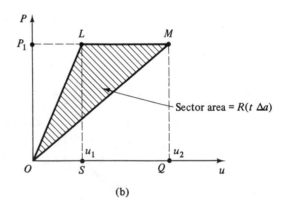

(b)

Figure 8-19 Same as Fig. 8-18 except crack propagates at constant load.

In terms of the designations in Fig. 8-19(b), this becomes:

$$(LS)SQ = \tfrac{1}{2}[(LS)OQ - (LS)OS] + R \,\Delta A \qquad (8\text{-}30)$$

or

$$R \,\Delta A = R(t \,\Delta a) = \tfrac{1}{2}(LS)SQ$$

which is again the energy represented by the sector area OLM. Note that in this situation the strain energy of the body *increases* as the crack propagates.

For a large number of solids, using properly designed test pieces, quasi-static cracking proceeds under conditions shown in Fig. 8-20; in essence,

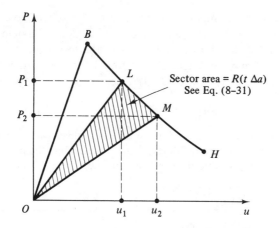

Figure 8-20 General P-u plot for quasi-static crack propagation showing how fracture toughness, R, is again related to a sector area.

this falls between the two cases just discussed. However, the fracture toughness, R, is again determined by associating the energy depicted by a sector area with the change ΔA that is taking place. In Fig. 8-20 consider area $OLMu_2O$. Let areas be designated as follows:

$$\text{I} = OLu_1, \quad \text{II} = OMu_2, \quad \text{III} = u_1LMu_2, \quad \text{and} \quad \text{IV} = OLM.$$

As the crack propagates and the load drops from P_1 to P_2,

$$\Delta A = \text{II} - \text{I}$$

and the external work is Area III. Using Eq. (8-29),

$$\text{III} = (\text{II} - \text{I}) + \text{IV}$$

so the shaded sector area, IV, is $R\,\Delta A$ as in the two earlier cases.

If during the load drop from L to M on Fig. 8-20, the increase in crack area is measured, R can be readily determined. A useful method in such data reduction is available in cases where the sector area may be reasonably described by the connection of straight lines, which is always possible if points L and M are made to approach each other. If there is large curvature to path $BLMH$, this may not be practical. Considering L as point one and M as two, the fracture toughness will be found from:

$$R = \frac{1}{2}\left(\frac{P_1u_2 - P_2u_1}{t\,\Delta a}\right) \qquad (8\text{-}31)^*$$

The proof of this result is left as an exercise.

If the calculation using Eq. (8-31) is felt to be of insufficient accuracy then the actual area of OLM can be determined (e.g., with a planimeter) and

*Robert Edstrom, a former student, first suggested this result.

properly converted to units of energy. One of the added advantages of the Gurney method is that a number of measures of R can be determined from a single test and if the crack *speed* varies as the test proceeds, any obvious rate effects on R may be noted. If it is found that regardless of the sector area used, R does not vary, then *BLMH* depicts a *constant R* locus.

Example 8-8.

The sketch illustrates a load-deflection curve obtained with a compact tension specimen used in fracture toughness studies. Determine the fracture toughness via the Gurney approach if the specimen thickness were 100 mm.

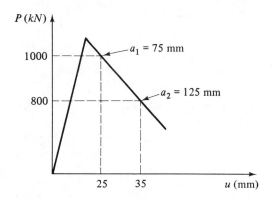

Solution.

From Eq. (8-31)

$$R = \left(\frac{1}{2}\right)\left(\frac{P_1 u_2 - P_2 u_1}{t\,\Delta a}\right)$$

where $P_1 = 1000$ kN, $P_2 = 800$ kN, $u_1 = 25$ mm, $u_2 = 35$ mm, and $\Delta a = 50$ mm.

$$R = \left(\frac{1}{2}\right)\left(\frac{1000(35) - 800(25)}{100(50)}\right) = 1.5\,\frac{\text{MJ}}{\text{m}^2}$$

Example 8-9.

Consider a split beam as shown in the sketch, where the crack length from the line of loading is 10 in., the specimen thickness is 0.250 in. and the height of the individual beams is 1 in.

(a) Based upon the Gurney approach, develop an expression for the load P necessary to cause crack propagation.

(b) If the yield strength of the material is Y, determine the conditions that lead to gross yielding before crack propagation.

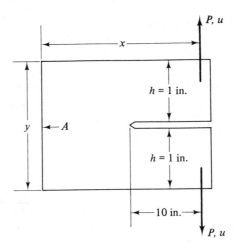

Solution.

(a) The total displacement is $u_1 + u_2$ and because both beams are equivalent in all respects, the total displacement becomes:

$$u = \frac{2Pl^3}{3EI} \qquad \text{or} \qquad \frac{u}{P} = \frac{2l^3}{3EI} = \frac{2A^3}{3EIt^3}$$

since the crack area, A, equals lt. Then,

$$\frac{d}{dA}\left(\frac{u}{p}\right) = \frac{2A^2}{EIt^3} = \frac{2l^2t^2}{E(\frac{1}{12})(t)(1)(t^3)} = \frac{24l^2}{Et^2}, \quad \text{since } h^3 = 1.$$

Now*

$$P^2 = \frac{2R}{\dfrac{d}{dA}\left(\dfrac{u}{p}\right)} = \frac{2REt^2}{24l^2} = \frac{RE}{19,200} \qquad \text{or} \qquad P \approx \frac{\sqrt{RE}}{139}$$

(b) The maximum bending stress in either beam comes from:

$$S = \frac{Mc}{I} = \frac{(\frac{1}{2})M}{\frac{1}{12}(\frac{1}{4})(1)} = 24M = 24(10P) = 240P$$

P to cause yielding equals $Y/240$ so if $Y/240 < \sqrt{RE}/139$ then yielding will precede cracking. Note too that if the dimension x is too small, a plastic hinge will develop due to compressive stresses in the vicinity of A. If the dimension y is too large, yielding through the net section may occur. Thus, the importance of specimen design becomes apparent if one intends that crack propagation occurs *before* gross yielding with the exception of a possible small plastic zone contiguous with the crack tip.

To generalize the previous cases and to introduce certain useful relationships, an energy balance may be written in differential form as follows:

$$P\,du = d\Lambda + R\,dA + d\text{Kin} \tag{8-32}$$

*See Eq. (8-36).

where the external work is equated to the strain energy, the fracture toughness work and kinetic energy. For the developments that follow, it will be assumed that:

1. Kinetic effects are negligible, therefore $d\text{Kin} \approx 0$.
2. Reversible behavior prevails both before and after the crack has propagated; thus, unloading the testpiece would always produce a return to the origin of the P-u plot.

Since strain energy may be expressed by $\frac{1}{2}Pu$,* Eq. (8-32) may be written as:

$$P\,du = d(\tfrac{1}{2}Pu) + R\,dA \tag{8-33}$$

which, upon expanding, rearranging, and dividing by P^2 gives:

$$\frac{du}{P} - \frac{u\,dP}{P^2} = \frac{2R}{P^2}\,dA \tag{8-34}$$

or

$$d\left(\frac{u}{P}\right) = \frac{2R}{P^2}\,dA \tag{8-35}$$

Thus,

$$P^2 = \frac{2R}{\dfrac{d}{dA}\left(\dfrac{u}{P}\right)} \tag{8-36}$$

Physically, the term $(d/dA)\,(u/P)$ represents the rate of change of compliance with respect to a change in crack area. Note the equivalence of Eq. (8-36) with Eq. (8-21).

From Eq. (8-33) some useful relations may be developed to include the following:

(a) If $P = $ constant, then

$$P = \frac{2R}{\left(\dfrac{\partial u}{\partial A}\right)_P} \quad \text{or} \quad R = \frac{\partial}{\partial A}\left(\tfrac{1}{2}uP\right)_P \tag{8-37}$$

(b) If $u = $ constant, then

$$u = \frac{-2R}{\left(\dfrac{\partial P}{\partial A}\right)_u} \quad \text{or} \quad R = -\frac{\partial}{\partial A}\left(\tfrac{1}{2}uP\right)_u \tag{8-38}$$

---

*Considering Figs. (8-19) and (8-20) as examples, the displacement related to external work (i.e. Pdu) is given as $u = (u_2 - u_1)$. The change in strain energy is really $\frac{1}{2}P(u_2 - u_1)$ or $\frac{1}{2}Pu$ and thus, $d\Lambda = \frac{1}{2}\,d(Pu)$. Problems 8-16 and 8-18 demand careful understanding of this point since unloading does not reverse back to the origin.

(c) For the general case,

$$u^2 = \frac{-2R}{\frac{d}{dA}\left(\frac{P}{u}\right)} \qquad \text{or} \qquad R = -\frac{1}{2}u^2\frac{d}{dA}\left(\frac{P}{u}\right) \tag{8-39}$$

where $(d/dA)\,(P/u)$ is the rate of change of stiffness with respect to a change in crack area.

The various forms of Eqs. (8-36) through (8-39) all relate P, u, and A as the crack extends. For the general case, they represent algebraic expressions of the line *BLMH* on Fig. 8-20 and convenience would really dictate which form is most useful.

8-9 CRACK STABILITY IN PHYSICAL TERMS

Stable cracking has been defined and approached from different viewpoints that are primarily mathematical [15] or physical [16] in interpretation. Although fast crack movements are a necessary condition for unstable crack propagation they are not fully sufficient in that connection; in essence, crack propagation can be fast yet stable.

It is helpful to consider a physical meaning of stability in connection with the two general types of testing machines used to determine values of R. Screw-driven machines are considered to be hard or stiff and are character-ized by a continual increase in displacement, u, even when a crack is spreading in the testpiece. This occurs because the machine crosshead is physically displaced so long as the driving screws are turned. Hydraulically driven machines are called soft and are characterized by a continual increase in the boundary force, P. Figure 8-21 portrays a constant R locus for a special

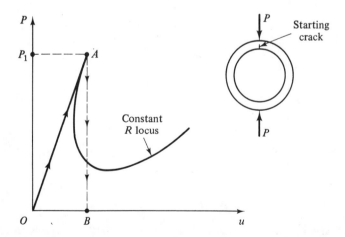

Figure 8-21 Theoretical constant R locus pertaining to a split ring speci-men.

testpiece such as a cracked ring which is indicated next to the plot. As the load P_1 is reached, the crack begins to lengthen and the R locus would begin a trace that starts back towards the origin. In this case, there is so much strain energy stored in the testpiece (OAB) that some could be removed immediately and there would still remain enough to feed the crack as it lengthens. To actually produce such a result as shown it would be necessary to reverse the screws of a hard machine as the crack spreads; in essence, the excess strain energy must be removed at a critical rate. Because most testing machines cannot accomplish this,* the availability of too much strain energy causes catastrophic failure and the load simply drops immediately from A to B. Thus the R locus would not be traced as shown and the situation describes instability.

With hard machines, stability requires that $du > 0$ as a constraint since positive crosshead displacement governs such devices. Note that on Fig. 8-21 du is negative since the R locus points back to the origin so an unstable condition prevails.

With soft machines, one must look to the behavior of P rather than u. Here, if P could be decreased at a critical rate as the crack begins to spread, again removing the excess strain energy, an unstable situation could be prevented. To do so in actuality is beyond the capabilities of soft machines and to produce stable propagation in this instance requires that $dP > 0$ in a mathematical sense.

Figure 8-22 should explain these two situations in a physical way. Any

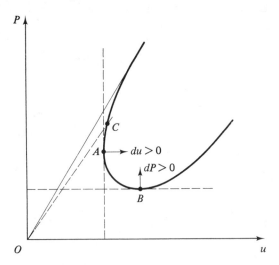

Figure 8-22 Physical meaning of crack stability.

*Electrohydraulic machines, such as those marketed by MTS, can act either hard or soft as desired. In addition, they may be programmed to trace the type of R locus shown on Fig. 8-21. Reference [17] can be consulted for details.

loading line to point A or to the right of A means that $du > 0$ as the crack begins to increase; thus by utilizing a specimen whose *starting* crack size produces an acceptable stiffness or compliance, stability may be achieved. Hard machines will therefore give stable crack propagation for any loading line that would *intersect* the R locus at a load less than P_A.

Since soft machines require $dP > 0$ for stability, point B describes the critical condition. A loading line that would intersect the R locus at B or to the right of B should lead to stable crack growth. As can be seen, hard machines provide a much broader range of P-u behavior that will produce stable situations as compared to soft machines. Because of this they are more satisfactory when conducting experiments to measure R. A loading line such as OC in Fig. 8-22 leads to instability regardless of the type of machine used. In the type of studies conducted by Griffith, the very small crack sizes used would, in effect, cause such loading behavior. For that reason, Griffith cracks were unstable as shown by his experiments.

Figure 8-23(a) illustrates the situations where both $du > 0$ and $dP > 0$ so stability would occur with either hard or soft machines. In Fig. 8-23(b), which is perhaps the most usual type of result, stability occurs with a hard machine since $du > 0$, but an unstable situation would be noted with a soft machine since $dP < 0$.

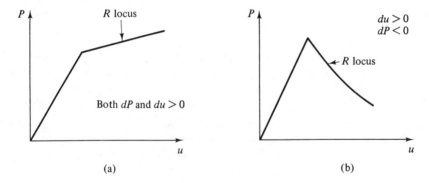

Figure 8-23 Two possible conditions related to stability using hard and soft testing machines.

8-10 CRACK STABILITY IN MATHEMATICAL TERMS

Considering first a hard machine where $du > 0$ for stable cracking, and using Eq. (8-39),

$$u^2 = -\frac{2R}{\dfrac{d}{dA}\left(\dfrac{P}{u}\right)}$$

which when differentiated with respect to A gives:

$$2u\frac{du}{dA} = \frac{-\frac{d}{dA}\left(\frac{P}{u}\right)2\frac{dR}{dA} + 2R\frac{d^2}{dA^2}\left(\frac{P}{u}\right)}{\left[\frac{d}{dA}\left(\frac{P}{u}\right)\right]^2} \tag{8-40}$$

or

$$2u\frac{du}{dA} = \frac{-2\frac{dR}{dA}}{\frac{d}{dA}\left(\frac{P}{u}\right)} + \frac{2R\frac{d^2}{dA^2}\left(\frac{P}{u}\right)}{\left[\frac{d}{dA}\left(\frac{P}{u}\right)\right]^2} \tag{8-41}$$

If each term in Eq. (8-41) is divided by u^2, which is equal to $-2R/(d/dA)(P/u)$, there results:

$$\frac{2}{dA}\frac{du}{u} = \frac{1}{R}\frac{dR}{dA} - \frac{\frac{d^2}{dA^2}\left(\frac{P}{u}\right)}{\frac{d}{dA}\left(\frac{P}{u}\right)} \tag{8-42}$$

With the imposed condition that $du > 0$ for stability, both u and dA must increase in a positive sense; thus the criterion for stability may be expressed as:

$$\frac{1}{R}\frac{dR}{dA} - \frac{\frac{d^2}{dA^2}\left(\frac{P}{u}\right)}{\frac{d}{dA}\left(\frac{P}{u}\right)} > 0 \tag{8-43}$$

The first term in Eq. (8-43) can be viewed as indicating any rate dependency of R with increasing crack area or length since $dR/dA = (dR/d\dot{A})(d\dot{A}/dA)$ whereas the second term is considered to be a geometric stability factor (g.s.f.) since P/u is dependent upon specimen geometry. Gurney and Mai [16] introduced the term geometric stability factor (g.s.f.).

Following an identical approach for soft machines, where $dP > 0$ must be imposed, and using Eq. (8-36) as a starting point, the analogous expression for the criterion of stability is found to be:

$$\frac{1}{R}\frac{dR}{dA} - \frac{\frac{d^2}{dA^2}\left(\frac{u}{P}\right)}{\frac{d}{dA}\left(\frac{u}{P}\right)} > 0 \tag{8-44}$$

With the use of either Eqs. (8-43) or (8-44) it is possible to predict if stability will likely occur for a particular specimen shape and type of machine to be used. What is demanded is the ability to compute either P/u or u/P on the basis of beam equations, elasticity theory, or whatever analysis is most appropriate. Even if exact solutions are unavailable, it is possible to attain a reasonable estimate of stability in this manner.

Example 8-10.

A material, whose fracture toughness, R, is constant, is produced in the specimen form shown. It is of thickness t, and its elastic modulus is E. If it is loaded as indicated and does not yield at points A, determine the condition for crack stability if a hard machine is used.

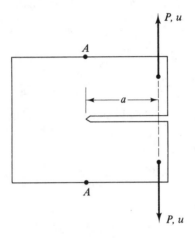

Solution.

Using Eq. (8-43) and noting that R is constant, stability demands that:

$$\frac{\dfrac{-d^2}{dA^2}\left(\dfrac{P}{u}\right)}{\dfrac{d}{dA}\left(\dfrac{P}{u}\right)} > 0 \qquad \text{where } dA = t\, da \text{ for constant } t$$

Treating each half of the split section as a cantilever beam and noting that the displacement per beam is half the full displacement (i.e., $u/2$):

$$\frac{P}{u} = \frac{3EI}{2a^3} = \frac{3EI}{2(A/t)^3} = \frac{3EIt^3}{2A^3}$$

$$\frac{d}{dA}\left(\frac{P}{u}\right) = -\frac{9EIt^3}{2A^4} \qquad \text{and} \qquad \frac{d^2}{dA^2}\left(\frac{P}{u}\right) = +\frac{36EIt^3}{2A^5}$$

$$\frac{+\dfrac{36EIt^3}{2A^5}}{-\dfrac{9EIt^3}{2A^4}} = -\frac{4}{A} = -\frac{4}{ta}$$

from the requirement for stability, $-[-4/(ta)] = 4/(ta) > 0$ since a and t are both positive. Thus, this test on a hard machine should lead to stable crack growth.

8-11 INITIAL CRACK VELOCITIES

If interest centers around the magnitude of crack speeds during the early stages of crack propagation, the following analysis is helpful.* Equation (8-33) may be written as:

$$P \, du = R \, dA + \tfrac{1}{2}(P \, du + u \, dP) \tag{8-45}$$

or

$$P \, du - u \, dP = 2R \, dA \tag{8-46}$$

This can be expressed as:

$$P - u\frac{dP}{du} = 2R\frac{dA}{du} = 2R\frac{\dot{A}}{\dot{u}} \tag{8-47}$$

where $\dot{u}$ is the machine crosshead velocity and $\dot{A}$ relates to the crack velocity. For a constant thickness specimen, $\dot{A} = t\dot{l}$ where t is the thickness and $\dot{l}$ is crack velocity. Noting that

$$u\frac{dP}{du} = u\frac{dP}{dA}\frac{dA}{du} = u\frac{dP}{dA}\frac{\dot{A}}{\dot{u}} \tag{8-48}$$

Eq. (8-47) becomes:

$$P - u\frac{dP}{dA}\frac{\dot{A}}{\dot{u}} = 2R\frac{\dot{A}}{\dot{u}} \tag{8-49}$$

or

$$\frac{\dot{A}}{\dot{u}} = \frac{P}{2R + u\dfrac{dP}{dA}} \tag{8-50}$$

An equivalent expression for constant thickness is,

$$\frac{\dot{l}}{\dot{u}} = \frac{1}{t}\left(\frac{P}{2R + u\dfrac{dP}{dA}}\right) \tag{8-51}$$

If $u \, (dP/dA)$ can be found from analysis, a prediction of the approximate starting crack velocity, $\dot{l}$, can be made. It should be noted that if an unstable R locus (see Fig. 8-21) could be traced, Eq. (8-51) would give a negative value for $(\dot{l}/\dot{u})$ during the unstable portion of the locus; this results since du is negative in that region.

By way of a summary, the following should be noted:

1. In comparing Eqs. (8-21) and (8-36), the terms $(1/M)$ and (u/P) are equivalent so that Gurney's fracture toughness, R, and the critical strain energy release rate, G_c are in reality identical, for the *start* of crack propagation.

*Atkins and Caddell [18] first presented this analysis.

2. From Eq. (8-26), as an example, $K_c^2 = EG_c = ER$ so for a particular specimen geometry and starting crack size, a measure of R could lead directly to a proper value of K_c if *plane stress* conditions were satisfied.*

3. From Eq. (8-36):

$$P^2 = \frac{2R}{\frac{d}{dA}(u/P)} \qquad (8\text{-}52)$$

and with $R = K^2/E$

$$P^2 = \frac{2K^2}{E \frac{d}{dA}(u/P)} \qquad (8\text{-}53)$$

Thus the influence of the $d/dA(u/P)$ term has, in effect, been introduced via the calculation of K in the stress calculations using numerous physical situations.

4. From the general expressions that

$$K = C\sigma(a)^{1/2} \qquad \text{and} \qquad R = \frac{1}{2}P^2 \frac{d}{dA}\left(\frac{u}{P}\right)$$

the stress intensity approach requires an analysis for the proper value of C for a given configuration whereas Gurney's approach requires an analysis to give the correct $d/dA\ (u/P)$.

8-12 INFLUENCE OF SPECIMEN SIZE ON FRACTURE TOUGHNESS MEASUREMENTS

Because the effects of plane stress versus plane strain behavior have such an important bearing on the values of fracture toughness obtained from experiments, it is crucial to add further comments in connection with Figs. 8-12, 8-14, and 8-15.

From Fig. 8-12 it was noted that the value of σ_y, whose distribution was predicted from elasticity, must cut off when it reaches the yield strength Y of the material; at that point plastic deformation begins. A plastic zone would form ahead of the crack tip and the radius of the zone, r_p, may be estimated as follows:

$$Y = K(2\pi r)^{-1/2} \qquad (8\text{-}54)$$

so,

$$r_p = \frac{1}{2\pi}\left(\frac{K}{Y}\right)^2 \qquad (8\text{-}55)$$

*If the term $(1 - v^2)$ is considered as in Eq. (8-27) then K_{Ic} can be found from R assuming plane strain conditions exist.

for conditions of plane stress. If plane strain prevails, the plastic zone radius has been estimated as:

$$r_p = \frac{1}{6\pi}\left(\frac{K}{Y}\right)^2 \qquad (8\text{-}56)$$

thus implying a lesser degree of plastic deformation. Figure 8-24 illustrates this in a schematic way where plane stress prevails at the free surface while the interior experiences plane strain.* Under such conditions, *shear lips*

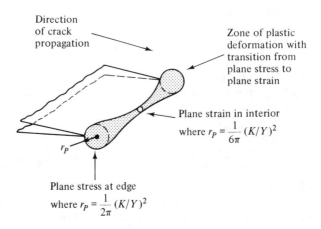

Figure 8-24 Illustration of plastic zone at a crack tip showing effects of plane stress (surface) and plane strain (interior).

result at the outer part of the fracture surface while a more brittle behavior is displayed as a flat region in the interior; Fig. 8-15 will assist here. The key physical point is that plane strain involves a lesser degree of plastic deformation than does plane stress and, as a consequence, the energy required to cause *crack propagation* is lessened as the extent of plane strain behavior is increased. Thus, experiments conducted under plane strain conditions produce lower or more conservative values of fracture toughness when compared with those where the plane stress contribution is more dominant. This is illustrated in Fig. 8-14. Much effort has been devoted to this point of concern and the result has led to standardized test specimens which, hopefully, produce plane strain values of fracture toughness.

Rather than simply including a sketch of the two standard specimens and their important dimensions, an explanation of the possible effects of specimen size on experimental measurements seems essential for a fuller

---

*The zone shape is also a function of θ. Experiments by Hahn and Rosenfield [19] show this clearly.

understanding. The crucial point of concern is related to the desire that no *yielding* should occur anywhere in the test specimen *except* for the possible existence of a small plastic zone contiguous with the crack tip!

If the thickness of the test specimen, B,* were on the order of twice r_p, then for plane stress conditions, Eq. (8-55) gives:

$$B = 2\left(\frac{1}{2\pi}\right)\left(\frac{K}{Y}\right)^2 = \frac{1}{\pi}\left(\frac{K}{Y}\right)^2 \qquad (8\text{-}57)$$

and the formation of plastic zones from each free surface would penetrate to the center of the specimen. For all practical purposes, one would see shear lips across the section with little if any flat surface since plane strain effects would be negligible. A high value of fracture toughness would result due to the large degree of plastic deformation prior to crack propagation. To preclude this result, the minimum value of B has been standardized as:

$$B \geq 2.5\left(\frac{K}{Y}\right)^2 \qquad (8\text{-}58)$$

Note that with this restriction, B is about 5π times r_p under plane stress and $15\pi r_p$ under plane strain. What this indicates is that B is on the order of 47 times r_p if plane strain prevails so the major portion of the fracture surface will be flat and the type of failure shown in Fig. 8-15 would occur.

What is not fully apparent is the possibility of generalized yielding occurring in sections of the specimen *remote* from the crack tip prior to crack propagation. Such an occurrence, or its prevention, is also related directly to the specimen shape and size. Realize that if such yielding did occur prior to crack propagation, measured values of fracture toughness would also be high for the same reasons discussed above. Several possibilities that involve yielding away from the crack tip will now be discussed.

Reference to Example 8-9 indicates one source of yielding. The double cantilever beam specimen† could yield before cracking if certain dimensions are not properly controlled. From Example 8-9

$$P^2 = \frac{2R}{\dfrac{d}{dA}\left(\dfrac{u}{P}\right)} = \frac{2REIB}{2l^2} \qquad \text{or} \qquad M^2 = REIB \qquad (8\text{-}59)$$

since M equals Pl. The maximum fiber stress, $(MH)/(2I)$, will induce yielding of the beams when

---

*B is now commonly used to denote specimen thickness.

†This configuration, with controlled dimensions, is called a compact tension specimen (CTS).

$$\frac{MH}{2I} > Y \quad \text{or} \quad \frac{6M}{BH^2} > Y \tag{8-60}$$

Inserting Eq. (8-59) into Eq. (8-60) produces the result that yielding will precede cracking if

$$\frac{3RE}{H} > Y^2 \tag{8-61}$$

Thus, to preclude yielding:

$$H > \frac{3RE}{Y^2} \quad \text{or} \quad 3\left(\frac{K}{Y}\right)^2 \tag{8-62}$$

Note that Eq. (8-62) is independent of thickness. In terms of standardized nomenclature, H is defined as $1.2B^*$ which is exactly what is shown by Eqs. (8-58) and (8-62).

Another manner in which yielding can occur remote from the crack tip is due to the formation of a compressive plastic hinge on the face of the specimen furthest from the crack tip. This will occur if the length of material, into which the crack would propagate, is too small. In standard symbolization this is defined by a dimension W which is equal to $2a$.

Finally, the possibility of yielding in the net section (the sound cross-section in line with the starting crack) must be considered. A combination of tall specimens, i.e., large H, and unduly long starting cracks could produce such a result.

Because of these conceivable situations, one cannot be too arbitrary in regard to testpiece dimensions if valid fracture toughness measurements are desired. Figure 8-25 shows the main essentials in regard to the dimensions of a CTS. For complete details as well as a full description of the standardized test specimen the text by Rolfe and Barsom [7] can be consulted.

Figure 8-26 is a composite that indicates the possible causes of generalized yielding due to improper dimensions of test specimens and should assist in interpreting the previous remarks related to yielding before cracking. Suppose specimen *abcd* is dimensioned such that satisfactory measurements of fracture toughness would be determined. If specimen *aefd* were used, the possibility of yielding at U, prior to crack propagation, would be enhanced since a compressive plastic hinge is likely to occur. A specimen in the shape of *ghij* could lead to yielding across section WX or ZX, as in Example 8-9, because the height is too small. Finally, specimen *klmn* could experience gross yielding through section XU because of the combination of too large a height and starting crack length.

---

*With a CTS, both crack length a, and thickness B, are specified as being ≥ 2.5 $(K_{Ic}/Y)^2$. H is taken as $1.2a$. Note in Eq. (8-59) the moment arm, l, is essentially the same as a.

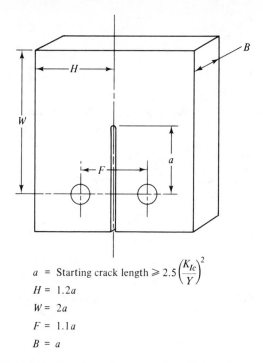

a = Starting crack length $\geqslant 2.5\left(\dfrac{K_{Ic}}{Y}\right)^{2}$

$H = 1.2a$

$W = 2a$

$F = 1.1a$

$B = a$

Figure 8-25 Principal dimensions of a compact tension specimen (CTS) [5].

Figure 8-26 Comparative specimen sizes that relate to the possibility of plastic deformation during fracture toughness testing.

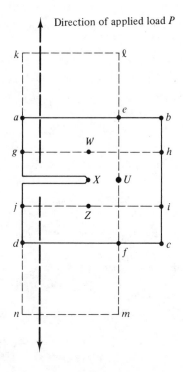

REFERENCES

[1] A. A. GRIFFITH, "The Phenomena of Rupture and Flow in Solids," *Phil. Trans. Roy. Soc. (London)*, **A221** (1920), pp. 163–98.

[2] C. E. INGLIS, "Stresses in a Plate Due to the Presence of Cracks and Sharp Corners," *Proc. Inst. Nav. Arch.*, **55** (1913), pp. 219–30.

[3] E. OROWAN, "Fracture and Strength of Solids," *Rep. Prog. in Phys. Soc. (London)*, **12** (1949), pp. 185–232.

[4] G. R. IRWIN, *Fracturing of Metals* (Cleveland, Ohio: ASM, 1949), pp. 147–66.

[5] "Standard Test Method for Plane-Strain Fracture Toughness of Metallic Materials," *ASTM Designation E399–74*, Part 10, ASTM Annual Standards 1977.

[6] H. M. WESTERGAARD, "Bearing Pressures and Cracks," *Trans. ASME, Jour. Appl. Mech.*, **61** (1939), pp. A49–A53.

[7] S. T. ROLFE and J. M. BARSOM, *Fracture and Fatigue Control in Structures* (Englewood Cliffs, N. J.: Prentice-Hall, Inc., 1977) pp. 32, 52–90.

[8] P. C. PARIS and G. C. SIH, "Stress Analysis of Cracks," in *Fracture Toughness Testing and Its Applications, ASTM STP 381* (ASTM: Philadelphia, 1965), pp. 30–81.

[9] H. TADA, P. C. PARIS, and G. R. IRWIN, *Stress Analysis of Cracks Handbook* (Hellertown, Pennsylvania: Del Research Corporation, 1973).

[10] G. C. SIH, *Handbook of Stress Intensity Factors* (Bethlehem, Pennsylvania: Lehigh University, 1973).

[11] D. P. ROOKE and D. J. CARTWRIGHT, *Compendium of Stress Intensity Factors* (Uxbridge, England: The Hillingdon Press, 1976).

[12] R. W. HERTZBERG, *Deformation and Fracture Mechanics of Engineering Materials* (New York: John Wiley & Sons, Inc. 1976).

[13] C. GURNEY and J. HUNT, "Quasi-Static Crack Propagation," *Proc. Roy. Soc., London*, **A299** (1967), pp. 508–24.

[14] C. GURNEY and K. M. Ngan, "Quasi-Static Crack Propagation in Nonlinear Structures," *Proc. Roy. Soc., London*, **A325** (1971), pp. 207–22.

[15] D. P. CLAUSING, "Crack Stability in Linear Elastic Fracture Mechanics," *Int. Jour. Fract. Mech.*, **5**, 3 (1969), pp. 211–27.

[16] C. GURNEY and Y. W. MAI, "Stability of Cracking," *Engr. Fract. Mech.*, **4** (1972), pp. 853–63.

[17] J. A. HUDSON, S. L. CROUCH, and C. FAIRHURST, "Soft, Stiff and Servo Controlled Testing Machines," reprint supplied by MTS Systems Corporation, Box 24012, Minneapolis, MN, 55424.

[18] A. G. ATKINS and R. M. CADDELL, presented at Winter Annual Meeting, ASME, New York, 1972.

244

[19] G. T. HAHN and A. R. ROSENFIELD, "Local Yielding and Extension of a Crack Under Plane Stress," *Acta Met*, **13** (1965), pp. 293–306.

[20] Y. W. MAI, A. G. ATKINS, and R. M. CADDELL, "On the Stability of Cracking in Tapered DCB Testpieces," *Int. Jour. Fract.*, **11**, 6 (1975), pp. 939–53.

PROBLEMS

8-1 When normally brittle materials are tested in torsion they might yield instead of cracking. Explain why with the use of Mohr's circle.

8-2 A class 20 gray cast iron has an average graphite flake size about 9 times as long as a class 60 type. The strength at fracture of the former is about 20,000 psi. Using arguments that would follow Griffith's analysis, what would you predict would be the strength of the class 60 iron?

8-3 A cylindrical pressure vessel, having a diameter of 20 ft and wall thickness of 1 in. underwent catastrophic failure when the internal pressure reached 2500 psi. The vessel material had an elastic modulus of 30×10^6 psi, a yield strength of 350,000 psi, and a value of G_c of 750 lbf-in./in.^{2}.
   (*a*) Show that failure would not have been expected if the von Mises criterion had served for design purposes.
   (*b*) Based upon Griffith's analysis determine the size of crack that might have caused this failure. Example 8-2 may assist.

8-4 A portion of a gas valve made of gray cast iron is in the shape of a tube of 124 mm OD and 106 mm ID. If a radial crack 47 mm in the axial direction and 5 mm deep in the radial direction extends inward from the outer surface, estimate the K_{Ic} of this cast iron if it fractured under an internal pressure of 1.5 MPa. All dimensions are in mm.

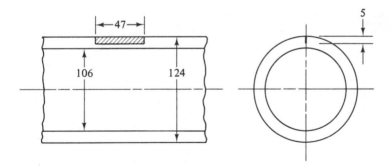

8-5 A wing panel for a supersonic airplane made of titanium alloy UNS R56401, of 1035 MPa yield strength, is 3.0 mm thick, 2.40 m long, and 2.40 m wide. The average cyclic tensile stress in the length direction is 700 MPa, insufficient for yielding but enough to cause gradual growth of a small pre-existing transverse flaw midway along one edge of the plate. If the flaw is initially 0.5 mm long and grows at a rate $da/dN = 120$ nm/cycle, where a is the edge crack length and N is the number of stress cycles, calculate the number of stress cycles prior to catastrophic fracture of the panel. K_{Ic} for this material is 55 MPa·$\sqrt{\mathrm{m}}$.

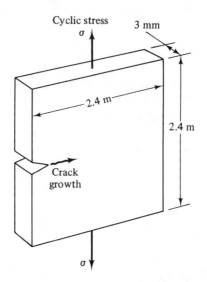

8-6 A flat plate of UNS A92024–T851 aluminum alloy contains a hole of 12 mm diameter. A second plate, shown in the sketch where all dimensions are in mm, is identical to the first except it also contains two 3-mm deep cracks in the hole. Find the ratio of the stress to cause fracture of the second plate to

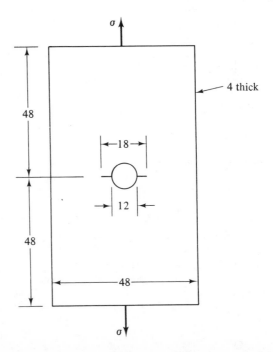

the stress to cause yielding of the first plate. The yield strength of this material is 455 MPa, K_{Ic} is 26 MPa $\cdot \sqrt{m}$, and the stress concentration factor, K_t, is 2.43. K_t is used in the usual context of macroscopic design.

8-7 A designer is considering T20811 steel for a high-strength part, to be loaded as shown in the sketch. The smallest edge crack that can be detected is 0.1 mm, and all dimensions shown are in mm. This material can be treated to produce different microstructures. In one case the structure exhibits a yield strength of 1790 MPa and K_{Ic} of 38 MPa $\cdot \sqrt{m}$; in the second condition, Y is 2070 MPa and K_{Ic} is 28 MPa $\cdot \sqrt{m}$.

    (*a*) Which T20811 steel should he use for minimum weight, the low-strength or the high-strength one?

    (*b*) Calculate the mass of the final plate if the density of this alloy is 8 Mg/m³ and the factor of safety is 1.6.

    (*c*) Calculate the mass of the plate if the other T20811 steel has been used instead, still using f.s. = 1.6.

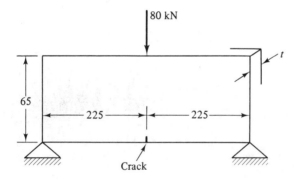

8-8 As a cracked plate is loaded, the crack begins to increase when a load of 120 lb is reached and the crack continues to lengthen under constant load. The plate has displaced about 0.100 in. just as the crack began to run; when the crack stopped increasing in length the displacement of the plate was 0.200 in. During this displacement the crack area increased by 5 in.².

    (*a*) Calculate the external work done during crack propagation.

    (*b*) Calculate *R*, using the Gurney approach.

8-9 A simply supported beam, loaded at the center, broke suddenly when the applied load reached 1000 lb; at that instant the maximum deflection was $\frac{1}{8}$ in. If the beam has a cross-sectional area of 0.750 in.², estimate the value of *R* for this material. Do you think this value of *R* is higher or lower than the value that would be found if this material cracked in a quasi-static manner?

8-10 You obtain some load/crack length/deflection information from a certain material as shown on page 248:

If the plate had a constant thickness of $\frac{1}{2}$ in., determine a reasonable estimate of *R*. The modulus of this material was 15×10^6 psi. Give the best estimate of K_{Ic} possible with these data.

Load (lbf)	Deflection (in.)	Crack length (in.)
900	0.2	1
600	0.3	2
450	0.48	3

8-11 The calibration bar technique is to be used to find an average value of G_c for a given solid. For a series of tests, all using specimens of 25 mm thickness, the following data were obtained. After determining G_c for several conditions, convert the above data to a form that will permit a plot of P versus crosshead displacement, u. Using this plot, determine several values of R using the Gurney approach to compare R and G_c.

Crack Length a mm	Compliance $1/M$(mm/N $\times$ 10^{-6})	Load P — (N)
51	71.4	2372
76	211	1495
102	457	1175
127	828	979
152	1370	837
178	2056	739

8-12 Consider a beam, split unsymmetrically as shown. If this material has a value of $R = 1000$ (in. · lbf)/in^2 and $E = 30 \times 10^6$ psi, at what load will the crack begin to lengthen?

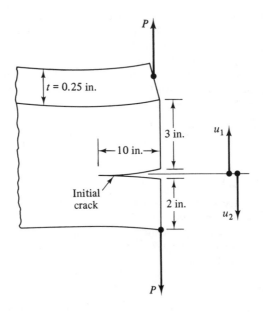

8-13 the Using relation $R = \frac{1}{2}P^2[\partial(u/P)]/(\partial A)$ and the graphical plot of a crack extending under constant load but with increasing crosshead displacement, it was shown that $R\,dA$ was equal to the "sector area" of a P-u plot as had been earlier shown by a purely graphical approach. Consider the case where a crack extends under constant displacement, u, but decreasing P. Starting with an R relationship that includes P, u, and A show that $R\,dA$ again equals the sector area on a P-u plot.

8-14 A tension specimen of thickness t and other dimensions as indicated is often used for fracture toughness studies. In such tests it is most desirable if the crack of initial length a propagates in a straight rather than a curved path. As a good first approximation, this seems probable if $\sigma_y > \sigma_x$ and if simple beam theory is acceptable. Determine relations for σ_x and σ_y as functions of P, a, W, and H. *Hint:* σ_y comes due to bending and direct tension whereas σ_x is due to bending alone. A detailed analysis of this problem is shown by Mai, et al. [20].

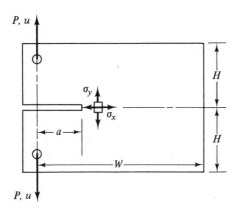

8-15 A split beam of thickness t, modulus E and starting crack length a, is subjected to loads as shown. Suppose this test is conducted on a hard machine and

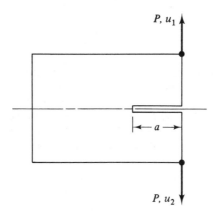

that R is independent of crack speed. Determine the condition for stability as a function of a. Repeat for a soft machine.

8-16   Calculate the force P required to remove masking tape of width w, thickness t, elastic modulus E and fracture toughness R between layers of tape, when a length L has been removed.

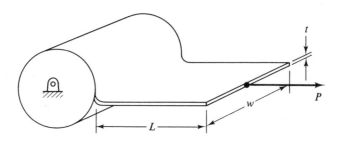

8-17   An elastic rod or fiber of diameter d is imbedded in a semi-infinite elastic body as shown. An initial crack of length L prevails and the fiber extends beyond the face of the medium by dimension x. As load P is increased, one of two things will eventually happen. Either the rod will break somewhere in the region of $L + x$ due to tensile failure or the initial crack will extend because the limit of fracture toughness R will be reached. Determine under what conditions cracking would occur along the interface, (i.e., L increases) rather than across the rod diameter. Note that pull-out of the fiber is not a factor in this analysis so the shear strength at the fiber-body interface is of no consequence here.

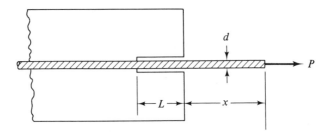

8-18   Tape, in a tapered shape as indicated, is being pulled from a drum by force P as shown. Assume the mass of the drum is of no consequence and deflections under load are negligible. If R is the fracture toughness at the interface between drum and tape and E is the elastic modulus of the tape, what force P is needed to remove the tape when L is the free length as shown? Note that the extension of the tape length L is PL/EA_{avg} where A_{avg} is the cross sectional area at $\frac{1}{2}L$ from the loaded end.

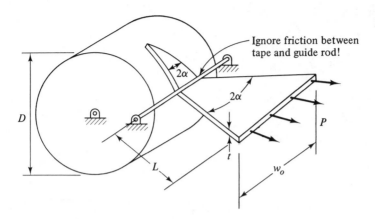

Ignore friction between
tape and guide rod!

8-19 Calculate the minimum force P required to split wood. Making some wild
estimates, let the initial configuration be as shown in the sketch below, and
further assume that homogeneous isotropic elastic beam theory holds (it
doesn't), that the elastic modulus of the wood is 14 GN/m², R is 5 kJ/m², and
the wedge is perfectly hard and rigid. Deflection δ of a cantilever beam
loaded at its end is $\delta = 4WL^3/Ebh^3$, where W is the load, L its length, E the
elastic modulus, b the base of the cross section and h its height. Frictional
effects are to be ignored.

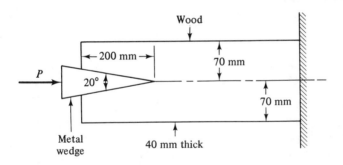

9

COMPOSITES

9-1 INTRODUCTION

From an historical viewpoint, composite materials have found practical use
for centuries. The use of straw in bricks and bull's hair in plaster typify man-
made composites used in ancient times.* During the past decade or so there
has been a tremendous increase of interest in composite materials for use in
engineering applications. Although the attractiveness of such structures stems
from a variety of reasons, the high strength to weight or stiffness to weight
ratios that are available, as compared with traditional bulk materials such as
metals, are, perhaps, the principal properties of greatest interest to engineers.

Many recent and current research studies, as well as actual industrial
applications, indicate that composites will find increasing use in the years
ahead. Although much of the past and present focus has been on aerospace
applications, it now seems probable that the use of such materials will find

---

*The delightful book by Gordon [1] contains an interesting discussion.

wider use in other fields. For that reason alone, design engineers should be aware of the potential of these interesting structures. As an example, the drive to lower the weight of automobiles, in order to improve their efficiency in regard to energy demands, should see a determined investigation of the possible use of composites as a replacement for metallic components. Since this chapter is only introductory in scope, certain restricted definitions and assumptions must be made in the discussion that follows.

9-2 DEFINITION OF COMPOSITES

For the remainder of this chapter, a composite carries the following attributes for the purpose of definition:

1. It is a man-made material; thus, certain natural structures such as wood, which contains sap, are excluded.
2. The individual materials that comprise the composite are bonded three-dimensionally in a continuous manner; thus, the laminated type of bimetallic strip used in thermostats is excluded.
3. The individual materials meet at a distinct interface and possess different chemical characteristics but do not rely upon any structural transformations during the formation of the composite; thus, the eutectic structure of many metallic alloys is excluded.

As defined above, these are sometimes called fibrous composites as contrasted with particulate and laminated composites.* If properly produced, composites can achieve a combination of properties that is superior to the properties of the individual constituents acting independently. Table 9-1,

TABLE 9-1*
Typical Unidirectional Material Properties

Material	Tensile Modulus (GPa)	Tensile Strength (GPa)	Density (Mg/m^3)
Fiberglass-epoxy	35	1.66	2.08
Boron-epoxy	207	1.38	2.02
Graphite-epoxy	138–276	1.38	1.58
Boron-aluminum	193	1.2	2.77
PRD-49-III	76	1.1	1.38
Aluminum (7075)	69	0.54	2.77
Ti-6A1-4V	110	1.17	4.43
Steel 4340	200	1.24	7.83

All original values converted to SI units.

*See Jones [2] for further details.

from a paper by Tang [3], provides a comparison of properties of some widely used composites and certain metals; this reference also discusses past and potential industrial uses.

Composites are often classified as being continuous, where the fibers or filaments span the entire length of the structure, or discontinuous, where the fiber length is less than the structure. These differences are analyzed below.

9-3   *CONTINUOUS FIBER COMPOSITES AND THE RULE OF MIXTURES*

Figure 9-1 illustrates a composite where the fibers are embedded in a matrix material and run the entire length of the composite. As an introduction, certain assumptions are made in regard to this particular type of composite. These are:

1. The fibers are parallel, unidirectional, and parallel to the intended loading direction.
2. The fibers are continuous.
3. Under applied loads, there is no relative movement at the fiber-matrix interface. With this assumption, the displacements of the fibers, matrix, and overall composite are identical during loading.

As the composite is subjected to external forces, both components will support a portion of the total load, where

L_f = the portion supported by all of the fibers

L_m = the portion supported by the matrix

L_c = the total load on the composite itself

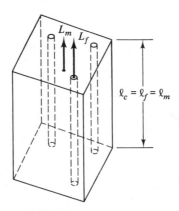

Figure 9-1 Section of a composite made with parallel and continuous fibers.

A simple force balance indicates:

$$L_c = L_f + L_m \tag{9-1}$$

or, in terms of the areas and induced stresses,

$$\sigma_c A_c = \sigma_f A_f + \sigma_m A_m \tag{9-2}$$

From the initial assumptions, equal lengths are involved, i.e., $l_c = l_f = l_m$, so the volume of the components is related directly to areas. It is common practice to talk of volume fractions, where

$$v_f = \text{volume fraction of all filaments}$$

$$v_m = \text{volume fraction of the matrix}$$

Thus, if the volume of the composite is taken as unity,

$$v_m = 1 - v_f \qquad \text{or} \qquad v_m + v_f = 1$$

With the above definitions, Eq. (9-2) may be expressed by:

$$\sigma_c = \sigma_f v_f + \sigma_m v_m = \sigma_f v_f + \sigma_m(1 - v_f) \tag{9-3}$$

This is usually called the Rule of Mixtures (ROM) and may be illustrated with the use of Fig. 9-2. Suppose the stress-strain behavior of the individual

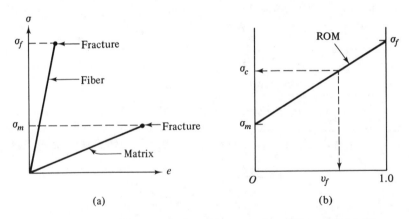

(a) (b)

Figure 9-2 Individual σ-e behavior of linear matrix and fiber materials (a) and method for relating composite strength (σ_c) to volume fraction of fibers (v_f) using the rule of mixtures (b).

constituents is given by Fig. 9-2(a) and it is further assumed that each constituent is capable of reaching its own fracture stress before the composite will fail. By plotting the individual fracture stress on the ordinates where $v_f = 0$ (i.e., no fibers), and $v_f = 1$ (i.e., no matrix), the connecting line describes the ROM. For a desired composite strength, σ_c, the intersection point defines the required value of v_f as shown in Fig. 9-2(b).

Besides strength requirements, stiffness, as defined by the elastic modulus, is also of great importance. In view of the assumption that there is no interfacial slip or movement, equivalent strains would be experienced, so

$$e_c = e_m = e_f \tag{9-4}$$

and with the relation that $\sigma = Ee$, Eq. (9-3) can be written as:

$$E_c = E_m v_m + E_f v_f = E_m(1 - v_f) + E_f v_f \tag{9-5}$$

It is important to realize that this expression for the modulus of the composite is independent of the fracture *stresses* of the individual components. Note also that if more than two components are involved, both Eqs. (9-3) and (9-5) would simply include additional but similar terms. Thus

$$\sigma_c = \sum_{i=1}^{n} \sigma_i v_i \quad \text{and} \quad E_c = \sum_{i=1}^{n} E_i v_i \tag{9-6}$$

Since the designer has, at least theoretically, the choice of selecting the constituents that make up a composite, an interesting trade-off between the proportion of the total load carried by the constituents and their respective elastic moduli can be shown. As implied by Eq. (9-4):

$$\sigma_m = \sigma_f \frac{E_m}{E_f} = \sigma_f \bar{E} \quad \text{where} \quad \bar{E} = \frac{E_m}{E_f} \tag{9-7}$$

Combining Eq. (9-7) with Eq. (9-3) gives:

$$\sigma_c = \sigma_f v_f + \sigma_f \bar{E}(1 - v_f) \tag{9-8}$$

or

$$\frac{\sigma_c}{\sigma_f} = \bar{E} + v_f(1 - \bar{E}) = \bar{\sigma}_c \tag{9-9}$$

noting that for a given $\bar{E}$, σ_m/σ_f is constant.

Figure 9-3 illustrates the variation of $\bar{\sigma}_c$ with v_f for two values of $\bar{E}$. As $\bar{E}$ decreases, E_m/E_f decreases and for a given value of v_f, the fibers must support a proportionally greater portion of the applied stress. If it is desirable

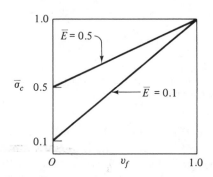

Figure 9-3 Effect of normalized modulus, $\bar{E}$, upon normalized composite strength, $\bar{\sigma}_c$, as a function of volume fraction.

to have the fibers carry a larger portion of the total applied load, then an increase in v_f or a decrease in $\bar{E}$ is appropriate.

Example 9-1.

The stress-strain curve of a matrix material fractures at a stress 0.2 GPa and a strain of 0.25 while the similar values for a filament are 2.0 GPa and 0.01 respectively. Both materials display linear behavior to fracture and the circular filaments are 0.1 mm in diameter. A composite of these two materials displays a strength of 0.8 GPa and is made according to the constraints listed in Sec. 9-3. Based upon the rule of mixtures:

(a) How many filaments are used if the area of the composite is 5 mm²?
(b) What is the elastic modulus of the composite?

Solution.

(a) With Eq. (9-3):

$$0.8 = 2v_f + 0.2\,(1 - v_f) \qquad \text{so} \qquad v_f = 0.33 = \frac{A_f}{A_c}$$

The total area of all filaments is 0.33(5) mm² so the number of filaments is $n = [0.33(5)]/[\pi(0.1)^2/4] = 210$ filaments.

(b) With Eq. (9-5):

$$E_c = \frac{0.2}{0.25}(1 - 0.33) + \frac{2.0}{0.01}(0.33) = 67.2 \text{ GPa}$$

9-4 THE MODIFIED RULE OF MIXTURES (MROM)

In regard to the strength of composites as predicted by the ROM, many experimental results show a decided discrepancy. As a step to account for this divergence, a modification was proposed by Kelly and Davies [4]. Their suggestion is referred to in this text as the modified rule of mixtures (MROM) and is based upon the concept that the largest strain to be tolerated by the composite when fracture is incipient is the smallest *fracture* strain of any individual component. Consider Fig. 9-4(a) where the stress-strain behavior of the fiber and matrix materials is superimposed on one plot. If fracture of the composite is governed by the smallest fracture strain of either component (in this case e_f of the fiber) then the largest *stress* ever experienced by the matrix would be σ'_m rather than σ_m as shown. In essence, the full stress capacity of the matrix cannot be utilized. Figure 9-4(b) indicates how the effect of σ_c versus v_f is predicted by the MROM. The dotted lines imply that as small values of v_f are introduced, the composite is "matrix controlled" and shows a strength that is *less than* the matrix strength alone. Obviously, this defeats one of the major purposes in producing composite materials and it is not until some critical volume fraction of fibers, v_{fc}, is introduced that the

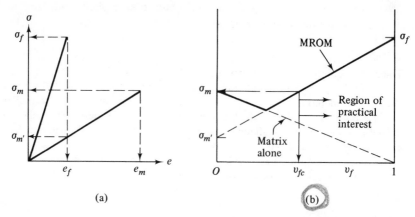

Figure 9-4 Same as Fig. 9-2 except the modified rule of mixtures is used.

composite strength exceeds the matrix strength. On these grounds, the region of practical interest starts at values of $v_f > v_{fc}$ as indicated.

With this as a constraint, the strength of a composite based upon the MROM is:

$$\sigma_c = \sigma_f v_f + \sigma'_m(1 - v_f) \qquad \text{where} \qquad v_f > v_{fc} \qquad (9\text{-}10)$$

Example 9-2.

Using the data in Example 9-1, what is the strength of the composite based upon the modified rule of mixtures? What is the elastic modulus for this case?

Solution.

When the filaments reach their fracture strain of 0.01, the matrix experiences a stress of 0.2/25 or 0.008 GPa.

Using Eq. (9-10):

$$\sigma_c = 2(\tfrac{1}{3}) + 0.008(\tfrac{2}{3}) = 0.672 \text{ GPa}$$

The elastic modulus would still be 67.2 GPa since this is strictly a function of v_f and the individual moduli.

9-5   *DISCONTINUOUS FIBER COMPOSITES*

With most modern composites, the filament material is the more expensive component and the use of continuous fibers could be questioned on the basis of economics. Although chopped or discontinuous fibers are often mixed with the matrix material in a random orientation, thereby producing a more isotropic composite in regard to properties, attention here will be restricted to those situations governed by the same assumptions stated in

Sec. 9-3. The only difference between the developments in this section and that in Sec. 9-3 arises because the fibers are not continuous.

For simplicity, we consider a single fiber of length l and diameter d embedded in a matrix; Fig. 9-5 illustrates this situation. As the composite is

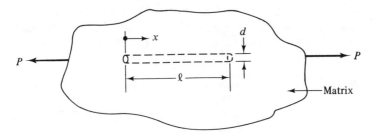

Figure 9-5 One discontinuous fiber in a matrix under load.

subjected to load P, the matrix away from the fiber must support the full load. In the interior region where the fiber is located, this component will sustain a transfer of the load from the matrix. The shift comes about due to the shear stress reaction at the interface; the term "shear stress transfer" is sometimes used in this context.

Figure 9-6 shows the stresses felt by the fiber when the ends of the composite are subjected to P. Considering the differential element shown, the

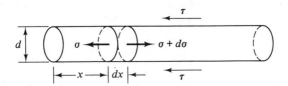

Figure 9-6 Elemental force balance on a fiber experiencing a shear stress at its periphery.

force balance would be:

$$\sigma\left(\frac{\pi d^2}{4}\right) + \tau(\pi d)\, dx - (\sigma + d\sigma)\frac{\pi d^2}{4} = 0 \tag{9-11}$$

Thus,

$$\frac{d\sigma}{dx} = \frac{4\tau}{d} \quad \text{and} \quad \sigma = \frac{4\tau x}{d} \tag{9-12}$$

if τ and d are considered to be constant. Thus, as x increases, the fiber experiences a larger stress with the limit being σ_f, the fracture stress of the fiber. The rate of change of the stress with x is given by $d\sigma/dx$; this may be called the transfer slope.

Figure 9-7 illustrates these comments. To the left of the point where x is zero, the matrix supports the entire load. As x increases, the matrix influence decreases whereas the fiber begins to support an ever-greater part of P. Where $x = \bar{x}$, the matrix support has dropped to zero and the fiber supports the entire load. As indicated on Fig. 9-7, the limiting value of σ is σ_f. Thus the shear stress transfer takes place over the length $\bar{x}$ after which the fiber carries the full load. It is of importance to investigate the influence of the fiber length l in this context, and several possibilities are now presented.

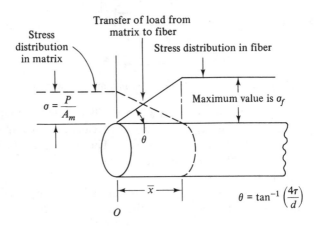

Figure 9-7 Transfer of load from the matrix to a discontinuous fiber according to the concept of the shear stress transfer.

CASE I: $\bar{x} = \frac{1}{2}l$

Here $\sigma = 2\tau l/d$ and if the full capacity of the fiber strength is to be attained then the stress should reach the value of σ_f. The value of fiber length to produce this condition is called the *critical fiber length* and denoted as l_c. Thus,

$$\sigma_f = \frac{2\tau l_c}{d} \tag{9-13}$$

and a useful design parameter is the critical-length-to-diameter ratio,

$$\frac{l_c}{d} = \frac{\sigma_f}{2\tau} \tag{9-14}$$

The ratio of σ_f/τ is on the order of 100; one often sees the ratio of 50 used as a design parameter for l_c/d. Naturally, not all composites have the same ratio of σ_f/τ so this general ratio of 50 must be viewed with some caution. Figure 9-8(a) shows the stress transfer for this case.

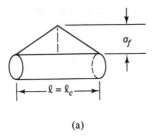

(a)

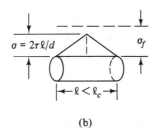

(b)

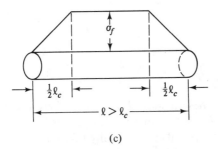

(c)

Figure 9-8 Shear stress transfer as affected by fiber length.

CASE II: $l < l_c$

Figure 9-8(b) illustrates this situation; here, the fiber experiences a stress less than σ_f so full advantage of its capabilities has not been taken.

CASE III: $l > l_c$

As shown in Fig. 9-8(c), the shear stress transfer occurs over a length $\bar{x}$ that is equal to $\frac{1}{2}l_c$; at that point, the maximum stress that can be tolerated is σ_f which is felt by the full central portion of the fiber.

9-6 CONCEPT OF THE AVERAGE FIBER STRESS

The above situations are often analyzed by considering the *average* fiber stress, $\bar{\sigma}$, as a measure of safety and because of somewhat better agreement with experimental results. Here

$$\bar{\sigma} = \frac{1}{l/2} \int_0^{l/2} \sigma \, dx = \frac{2}{l} \int_0^{l/2} \frac{4\tau x}{d} \, dx = \frac{\tau l}{d} \qquad (9\text{-}15)$$

Since $\bar{\sigma} < \sigma_f$, a comparison of Eqs. (9-13) and (9-15) indicates that if all other conditions are equivalent, longer fibers are required where the average fiber stress is used; this is the measure of safety mentioned above. For completeness, the three cases discussed earlier are now considered in terms of $\bar{\sigma}$ as the limiting stress.

CASE I: $l = l_c$

Figure 9-9(a) depicts this situation and from Eq. (9-15),

$$\bar{\sigma} = \frac{\tau l_c}{d}$$

thus, $\qquad\qquad\qquad\qquad\qquad\qquad\qquad\qquad\qquad\qquad\qquad (9\text{-}16)$

$$\bar{\sigma} = \tfrac{1}{2}\sigma_f$$

CASE II: $l < l_c$

This situation is described by Fig. 9-9(b) where the maximum stress felt by the fiber is $2\tau l/d$ and $\bar{\sigma}$ is simply $\tau l/d$ which is less than $\tfrac{1}{2}\sigma_f$. Again, in the sense of practicality, the fiber is not fully utilized.

CASE III: $l > l_c$

Figure 9-9(c) illustrates this condition and from Eq. (9-13) $l_c/2 = \sigma_f d/4\tau$. The average fiber stress is found by considering the areas in this figure as follows:

$$\bar{\sigma} l = 2\left(\frac{1}{2} \frac{l_c}{2} \sigma_f \right) + \sigma_f (l - l_c) \qquad (9\text{-}17)$$

There results:

$$\bar{\sigma} = \sigma_f \left[1 - \frac{l_c}{2l} \right] \qquad (9\text{-}18)$$

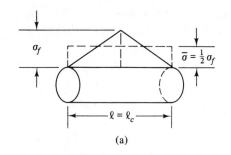

(a)

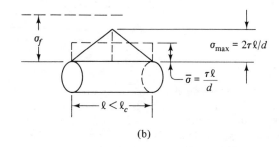

(b)

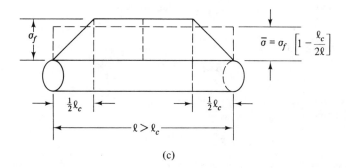

(c)

Figure 9-9 Effect of fiber length upon shear stress transfer using the concept of average fiber stress.

This finding has certain useful implications, these being,

1. If $l = l_c$, $\bar{\sigma} = \frac{1}{2}\sigma_f$ as in Fig. 9-9(a).
2. If $\frac{1}{2}l_c < l < l_c$, $\bar{\sigma} < \frac{1}{2}\sigma_f$ as in Fig. 9-9(b).
3. If $l < \frac{1}{2}l_c$ Eq. (9-18) is invalid and $\bar{\sigma}$ must be found from $\tau l/d$.
4. If $l > l_c$, $\bar{\sigma}$ will be greater than $\frac{1}{2}\sigma_f$. In this context it is more sensible to consider l_c as the *minimum* fiber length if the allowable fiber stress, σ_f, is to be used most advantageously.

In regard to the strength of such composites, this will of course depend upon the fiber length and will be illustrated using the ROM for simplicity. The basic relation is simply

$$\sigma_c = \bar{\sigma} v_f + \sigma_m v_m \qquad (9\text{-}19)$$

or

$$\sigma_c = \sigma_f \left(1 - \frac{l_c}{2l}\right) v_f + \sigma_m v_m \qquad (9\text{-}20)$$

Except for the case where $l < \frac{1}{2}l_c$, Eq. (9-20) is applicable, but as mentioned in the third comment below Eq. (9-18), that equation is invalid for such short fibers and $\bar{\sigma}$ is simply $\tau l/d$. Thus, where $l < \frac{1}{2}l_c$,

$$\sigma_c = \frac{\tau l}{d} v_f + \sigma_m v_m \qquad (9\text{-}21)$$

Example 9-3.

Discontinuous fibers are imbedded in a matrix and they lie parallel to the direction of intended loading. The fibers are 0.05 mm in diameter, possess a fracture strength of 3 GPa, and the interfacial shear stress between fibers and matrix displays a maximum value of 0.02 GPa. If the fibers are 2 mm long,
(a)   What is the maximum stress that will be felt by the fibers?
(b)   What is the average stress felt by the fibers?

Solution.

(a)   From Eq. (9-14),

$$l_c = \frac{\sigma_f d}{2\tau} = \frac{3(0.05)}{2(0.02)} = 3.75 \text{ mm}$$

so the fibers are less than the critical length.
The maximum stress experienced is:

$$\sigma = \frac{2\tau l}{d} = \frac{2(0.02)(2)}{(0.05)} = 1.6 \text{ GPa}$$

(b)   The average stress is, from Fig. 9-9(b),

$$\bar{\sigma} = \frac{\tau l}{d} = 0.8 \text{ GPa}$$

Because $l < l_c$ the full capability of the fibers is not used.

9-7   GENERAL COMMENTS

Several principles that may assist as general guides when composites are being considered for structural uses seem worth noting. These are given in terms of a general philosophy, since a misapplication in using composites as a substitute for other materials could be disastrous.

1. When fibers, whether continuous or discontinuous, are deliberately aligned, as discussed in Secs. 9-3 and 9-5, the resulting structure becomes anisotropic in terms of strength and stiffness. The use of anisotropic elasticity becomes imperative in regard to any analysis of properties.* Since the usual makeup of such composites involves fibers that are stronger and stiffer than the matrix, these properties are used most advantageously when the anticipated direction of maximum applied loads is parallel to the fibers. It would be foolish to subject such structures to the most severe loading in directions perpendicular to the fiber–matrix interface.

2. To develop isotropic composites, the fibers should be combined with the matrix in a random manner. Although the maximum strength will be bounded by the extreme values obtained with unidirectional fibers, its value will be little affected with regard to direction. Two such instances can be cited. If short or chopped filaments are introduced in a truly random manner, the composite will exhibit properties that approach complete isotropy. A lesser degree of isotropy results if the composite is built up of layers where the unidirectional fibers of each layer are positioned at various angles to adjacent layers.

3. Perhaps the single greatest conceptual failure with much composite design is to attempt to produce the strongest structure while giving little if any attention to toughness. As discussed in Chap. 8, the ability of a solid to withstand catastrophic crack propagation is often more critical than any other single characteristic. Only by possessing adequate fracture toughness can such a situation be controlled. As a general guide, as static strength increases, toughness decreases; so in "tailor-made" composites, a sacrifice of some degree of strength can be made to attain the necessary degree of toughness.

The concepts that follow provide some insight to procedures that improve toughness, but must of necessity be general and qualitative. As a start, consider the events shown in Fig. 9-10. In this extreme and simplified situation, assume that the interfacial strength is very high so a "strong" composite results.

When a certain load is reached, an existing crack begins to propagate in matrix material (in most cases, this is more likely as compared with initial propagation in the much smaller fibers). If the load is not reduced, the applied stress increases since a smaller sound area results as the crack extends. This leads to an unstable situation which, in the extreme cases of

---

*Jones [2] devotes about one hundred pages to this topic.

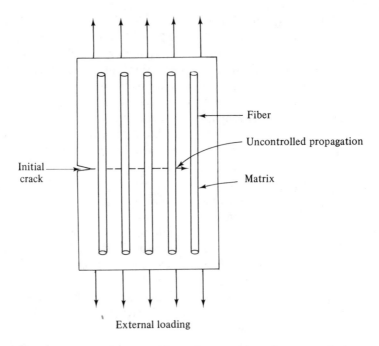

External loading

Figure 9-10 Type of catastrophic crack propagation when composite has inadequate fracture toughness.

brittle matrix and fiber, is basically described by Eq. (8-13) and Example 8-2. The crack length increases in a catastrophic manner and complete fracture of the cross section follows.

One of the earliest analyses regarding the influence of the relative adhesive strength at the fiber–matrix interface with the cohesive strength of the cross section of the solid was presented by Cook and Gordon [5]; an excellent and very readable coverage of the major concepts is given by Gordon [1].

The general details are shown in Fig. 9-11. Under applied stresses perpendicular to the crack, a tensile component of stress results ahead of the crack, and in the *plane* of the crack (or perpendicular to the applied loading). If this tensile stress exceeds the adhesive strength (i.e., the interface strength normal to the fibers), fracture can take place along the interface and *ahead* of the advancing front of the major crack. Two points are important. First, the creation of new surface area at the matrix–fiber interface must use up some of the available strain energy in the body and the new crack is parallel to the applied load; this is less serious than having the increase in crack area normal to the external load. Second, as the main crack blends with the interfacial break it is blunted and the main crack propagation can be arrested. By this mechanism, the excess strain energy can be sopped up by the creation of

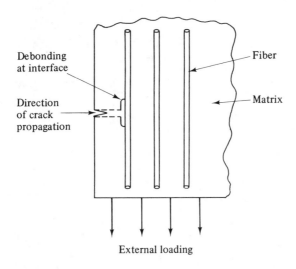

Debonding at interface

Direction of crack propagation

Fiber

Matrix

External loading

Figure 9-11 Cook and Gordon [1,5] debonding mechanism where the initial crack is blunted as it merges with the debonded region at the interface. This occurs *before* the crack reaches the fiber and debonding is caused by a tensile component in the plane of the crack ahead of the crack tip.

additional fracture along the interface and catastrophic failure thereby avoided. The key point resulting from this analysis is that by *decreasing* the interfacial strength, catastrophic failure may be avoided.

Using a boron-epoxy composite, with intermittent regions of low shear strength produced along the interface, Marston [6] showed that it was indeed possible to increase the fracture toughness as compared with the same composite that possessed a uniformly high strength at the interface. Figure 9-12 illustrates the manner by which a propagating crack may be stopped by this approach to composite design.

Further publications by Marston et al. [7] and Atkins [8] provide additional insight on this point. Felbeck [9] has extended the work of Marston [6] by using layers of composite materials having intermittent regions of low interfacial shear strength and reports a three-fold increase in fracture toughness with only a 26 percent decrease in static strength. In view of these results there is little question that a deeper understanding of the optimal combination of strength and toughness, for the many fiber–matrix combinations used as composites, is necessary before these materials are fully acceptable as replacements for the longer used and better understood metallic solids.

To provide support for the last comment, reference is made to the work of Outwater and Murphy [10] where fiberglass composites were used. In these composites, the fracture strain of the glass filaments is greater than

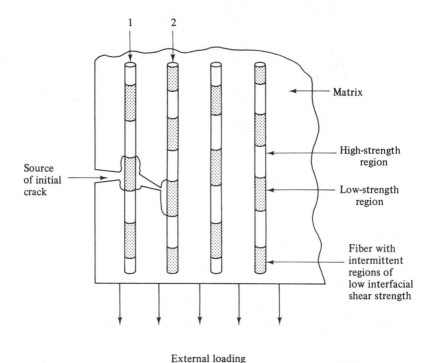

External loading

Figure 9-12 Fibers containing deliberately produced regions of low interfacial strength which are positioned in the staggered manner shown. Crack
propagation seeks the low-strength regions which debond ahead of the
crack. See plate 10 of Gordon [1] for an excellent photograph which shows
how interfaces can stop or hinder propagation.

that of the matrix. As a matrix crack reaches and passes around a fiber, the
fiber can still stretch without immediate failure. The added strain energy in
the fibers causes debonding along the interface; this continues until the fibers
themselves fracture. This implies that debonding at the interface does not
occur until the matrix crack has passed by a fiber. Figure 9-13 illustrates this
concept schematically.

Note that both debonding and fracture have occurred with fiber 1.
Although debonding has occurred at fiber 2, fracture has not yet occurred as
that fiber is still being stretched. Finally, debonding has not yet begun at
fiber 3 yet the crack has started to pass by this fiber.

The debonding mechanism illustrated by Fig. 9-13 is not consistent with
that associated with Fig. 9-11 since one proposes that debonding occurs
before the major crack reaches the fiber whereas the other requires the crack
to pass the fiber before debonding results. These points plus other factors

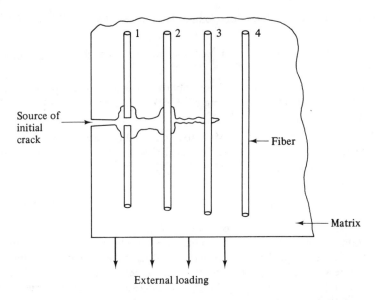

Figure 9-13 The Outwater-Murphy [10] mechanism of debonding. Here the crack passes a filament before debonding occurs.

that may contribute to improved toughness are given excellent coverage by Marston et al. [7]. At this stage it seems clear that different theories of the toughness of composites may be needed for the many combinations of matrix and fiber materials currently in use.

REFERENCES

[1] J. E. GORDON, *The New Science of Strong Materials or Why You Don't Fall Through the Floor*, (Harmondsworth, England: Penguin Books Limited, 1968).

[2] R. M. JONES, *Mechanics of Composite Materials* (New York: McGraw-Hill Book Company, 1975), pp. 2–10.

[3] S. TANG, "Advanced Composites: Design Technology Transfer from Aerospace," *Mech. Eng.* (June, 1977), pp. 36–39.

[4] A. KELLY and G. J. DAVIES, "The Principles of the Fiber Reinforcement of Metals," *Met. Rev.* (1965), pp. 1–77.

[5] J. COOK and J. E. GORDON, "A Mechanism for the Control of Crack Propagation in All-Brittle Systems," *Proc. Roy. Soc.* A282, (1964), pp. 508–520.

[6] T. U. MARSTON, Ph.D. dissertation (1973), The University of Michigan, Ann Arbor, Michigan.

[7] T. U. MARSTON, A. G. ATKINS, and D. K. FELBECK, "Interfacial Fracture

Energy and the Toughness of Composites," *Jour. Matl's. Sci.*, **9** (1974), pp. 447–55.

[8] A. G. ATKINS, "Intermittent Bonding for High Toughness/High Strength Composites," *Jour. Matl's Sci.*, **10**, (1975), pp. 819–32.

[9] D. K. FELBECK, "Fiber Reinforced Solids Possessing Great Fracture Toughness: The Role of Interfacial Strength," Final Tech. Report, NASA Grant No. NGR-23-005-528, July 31, 1977.

[10] J. O. OUTWATER and M. C. MURPHY, "On the Fracture Energy of Unidirectional Laminates," 24th Annual Technical Conference, Composites Division, *Soc. of Plast. Industry, Inc.*, Paper IIC (1969).

PROBLEMS

9-1 You wish to manufacture a composite tape made of a single layer of aligned continuous glass filaments of 0.005 in. diameter, which have a fracture strength of 100 ksi, embedded in a matrix of resin that has a strength of 5 ksi at the same fracture strain as the glass filaments. If the tape is to be 0.008 in. thick, 2 in. wide, and have a strength of 30 ksi, how many strands of glass filaments must it contain?

9-2 A matrix material has a linear stress strain curve that indicates a fracture stress of 50 ksi when the corresponding strain is 0.100. A filament material has similar values of 200 ksi and 0.010 respectively. Based upon the Modified Rule of Mixtures (MROM), what is the minimum v_f that should ever be used for practical considerations?

9-3 Two components, whose σ-e behavior is shown, are to be combined to form a composite whose strength is to be 75 ksi based upon the modified rule of mixtures (MROM). What would be the elastic modulus of the composite?

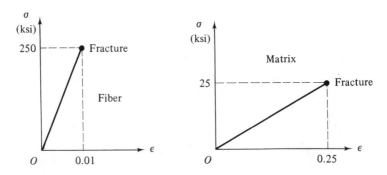

9-4 A matrix material shows a linear σ-e behavior where the fracture σ and e are 45 ksi and 0.150 respectively. Similarly, a filament material has a fracture σ and e of 400 ksi and 0.020 respectively.

  (*a*) From practical strength considerations, what is the *smallest* volume fraction of filaments to be used if the MROM is considered?

(b) What volume fraction of *matrix* would you use to design a composite having an elastic modulus of about 7 GPa (about one million psi)?

9-5 A composite is to be made from the materials whose individual stress-strain behavior is shown. If the composite is to reach a strain of 0.005 when the applied stress is 20 ksi, what volume fraction of fibers should be used? Show all calculations.

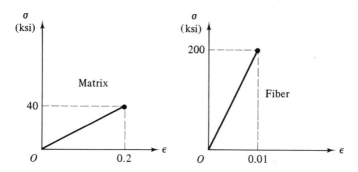

9-6 Tensile data on a continuous-filament glass fiber reinforced tape are as follows:

Test	Fracture Stress (ksi)	Orientation of Tensile Load
A	17.2	Longitudinal to fibers
B	2.9	45° to fibers
C	4.6	Transverse to fibers

Give a plausible explanation regarding why the variation of stress with orientation probably occurred.

9-7 Refer to Fig. 9-3 and consider a composite whose volume fraction of fibers is 0.3. For the two values of $\bar{E}$ shown, compare the stress felt by the fibers in each case.

9-8 Consider a composite containing discontinuous fibers positioned in a matrix as discussed in Example 9-3. The fibers have a fracture strength of 2 GPa, diameter of 0.04 mm, an interfacial shear strength of 0.01 GPa, and a length of 6 mm.

(a) Compare the fiber length with the critical length, l_c.

(b) What is the maximum stress transmitted to the fibers?

(c) What is the average stress felt by the fibers?

(d) If the strength of the matrix is about 0.02 GPa and the composite is to have a strength of 0.05 GPa, what value of v_f would be appropriate if the design is based upon the average stress carried by the fiber?

(e) If the value of v_f found in part (d) were used and the fibers were of critical length what would be the strength of the composite if the design was based upon the average fiber stress?

10

FATIGUE

10-1 INTRODUCTION

In numerous situations, components of a physical structure are subjected to stresses that fluctuate between particular levels. Although the maximum stress experienced is well below the tensile strength of the material and frequently below the *yield strength*, failure may occur after a sufficient number of cycles of stress fluctuation has transpired. This type of failure is called fatigue and even to this day, one may hear that "the metal has tired to the point of exhaustion" or, "the metal has crystallized and become too brittle." Such comments should not be made by educated engineers, or anyone else for that matter, since they are both incorrect and meaningless. After all, metals are crystalline to begin with and, unlike humans, do not possess the ability to feel tiredness.

It has been estimated that fatigue accounts for 90 percent of all component failures. Whether anyone has accurately documented this number is of little concern; the important point is that such failures are the most numerous type encountered and almost any design requires a consideration of this potential event. As in previous chapters, an introduction to important aspects will be covered to such a degree that further reading of more specialized sources can be pursued in a meaningful way.

At the outset, the most important factors that influence fatigue behavior are discussed, some only in a qualitative sense. Next, a brief coverage of the approach often used in mechanical design is given. In this context, the use of certain mechanical properties and available experimental information related to fatigue studies is introduced. A third aspect of this topic involves a more recent approach. This has to do with crack propagation and the concepts of fracture mechanics in regard to fatigue. From a truly fundamental viewpoint, the discussion of crack propagation provides the most meaningful explanation of *why* fatigue failures occur. Finally, a brief coverage of the visual analysis of fatigue failures is presented. Although this takes place after a failure has occurred, such a study often leads to an understanding of the causes of failure; this information may then be used to indicate how parts should be redesigned so that future failures will be avoided.

As a final introductory comment, there are two *necessary* conditions that must prevail if a fatigue failure is to occur. The initial prerequisite is the existence or initiation of a crack; this could be caused by the presence of inclusions, micro (or macro) porosity due to casting, defects due to welding, or as discussed briefly in Chap. 7, because of the behavior of dislocations.* For whatever the reason, the presence of a crack is essential to the fatigue process. The second prerequisite demands that the crack must propagate. Although many parameters may accelerate crack propagation, some type of fluctuating or alternating stress is essential to bring about a fatigue failure. Once a crack has propagated to some critical extent, the remaining sound section of the structure can no longer support the applied stresses and catastrophic failure follows. The regions of the longer term propagation and sudden failure show decidedly different appearances; this is discussed in the section on failure analysis.

10-2 FACTORS THAT INFLUENCE FATIGUE

STRESS FLUCTUATION

Several descriptive terms that describe the important components of stress are in common use; Fig. 10-1 illustrates possible situations where the following definitions pertain:

*Hertzberg [1] discusses other proposed models.

$$\sigma_m = \text{mean stress} = \frac{\sigma_{max} + \sigma_{min}}{2} \qquad (10\text{-}1)$$

$$\sigma_a = \text{stress amplitude} = \frac{\sigma_{max} - \sigma_{min}}{2} \qquad (10\text{-}2)$$

$$\sigma_r = \text{stress range} = \sigma_{max} - \sigma_{min} \qquad (10\text{-}3)$$

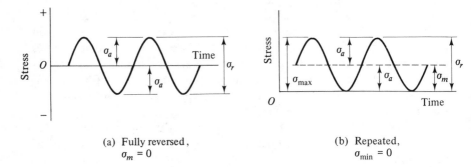

(a) Fully reversed, $\sigma_m = 0$ (b) Repeated, $\sigma_{min} = 0$

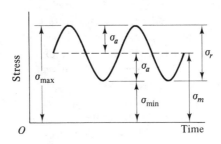

(c) Fluctuating

Figure 10-1 Several examples of cyclic stress conditions indicating pertinent stress parameters.

Although a sine curve is used for illustrative purposes, most real situations would not display such behavior. In any event, the shape of such a stress-time plot appears to have no particular significance; it is the fluctuating behavior that is of concern.

Since crack propagation is most susceptible to tensile stresses, any increase in σ_m (i.e., more tensile) will be detrimental and failure by fatigue will occur in a shorter time. Because maximum tensile stresses usually prevail at some surface of a stressed component, it should not be surprising that, with few exceptions, the path of crack propagation is from the surface to the inner portion of the particular cross section under stress.

SURFACE FINISH

Various texts discuss the influence of surface finish from two points of view. One considers this parameter to be a type of stress concentration on a micro scale, that is, machining marks, scratches, surface scale due to hot working, etc., are viewed as sources of stress concentration. Other books, however, distinguish between these surface variables and the macroscopic sources of stress concentration such as holes, fillets, keyways, etc.; this approach is followed here.

Figure 10-2 illustrates the influence of surface finish on fatigue via a modifying factor. The use of the factor C_s is discussed briefly in the Sec.

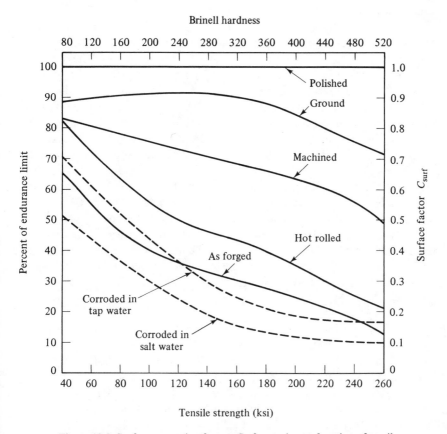

Figure 10-2 Surface correction factor, C_s, for steels as a function of tensile strength (reproduced with the kind permission of C. Lipson and R. C. Juvinall (eds.) from "Application of Stress Analysis to Design and Metallurgy," The University of Michigan Summer Conference, Ann Arbor, MI., 1961).

10-3 on mechanical design; for the time being it will suffice to state that smoother surface finishes are beneficial in prolonging fatigue failure where all other factors of influence are equal.

STRESS CONCENTRATION

The equations usually employed to calculate the magnitude of stresses induced in a body under load do not in themselves account for the influence of macroscopic section irregularities such as holes, fillets, sharp corners, and the like. In these regions, the actual stress state can be markedly different as compared to sections where no irregularities exist. Considering that the applied forces must be transmitted in a more concentrated manner, the alteration in the stress distribution in such regions is described as a *stress concentration*. The discontinuities themselves are called *stress raisers* since they raise the level of stress above computed nominal values. The influence of stress raisers is most often described by an elastic stress concentration factor, K_t. Note that this is entirely different in meaning from the stress *intensity* factor, K_c, discussed in Chap. 8. Numerous texts provide plots of K_t for many common geometries; Peterson [2] is an excellent source. The use of K_t by itself simply indicates the relative magnitude of the maximum stress caused by the discontinuity as compared to the nominal applied stress that is computed from standard equations (e.g., in bending, Mc/I). Many test results have indicated the need to reduce the apparent influence of K_t where fatigue is of concern; this has led to the concept of a *fatigue stress concentration factor*, K_f. Physically, K_f accounts for the lack of full sensitivity of K_t as to its influence upon the fatigue behavior of various materials. A tenable viewpoint is that K_f adjusts the influence of K_t to account for plastic flow, and K_f is often viewed as a *fatigue strength reduction factor*. To incorporate the two parameters, a term called the *notch sensitivity*, q, is most often used, where:

$$q = \frac{K_f - 1}{K_t - 1} \tag{10-4}$$

In the extremes, if $q = 1$, $K_f = K_t$ and the material is highly sensitive to stress raisers as to their influence on fatigue. If $q = 0$, $K_f = 1$ and the presence of stress raisers has no influence upon fatigue behavior as far as the effect of stress concentration is concerned. Although numerous values of K_t have been determined on analytical grounds and verified by experiment, values of K_f are determined by experiment alone and are found by taking the ratio of the *endurance limit*, discussed in Sec. 10-3, of notch-free to notched specimens. For a given combination of material response to K_f and various values of K_t caused by different stress raisers, the combined influence upon

q can then be determined. There is a subtle point here that is sometimes over-looked. If, for example, a shaft of two different section sizes possesses a sharp corner where the sections blend, it will have a higher value, of K_t compared with the same situation where a generous fillet joins the two sections. The magnitude of K_f would also differ because of the difference in the notch effect and the two values of q would then reflect both the influence of K_t and K_f. In essence, these parameters are not mutually independent. In real life designs it is always beneficial to keep K_t as small as possible and the value of K_f for a particular value of K_t then describes q. It should be realized that as the ability of a material to undergo plastic deformation increases, the influence of q decreases. Whenever values of K_t and q are required to solve problems in this text, they will be given directly. Juvinall [3] discusses this in detail and gives several examples showing that the undesirable effects of stress raisers can be lessened by *removing* material from the part.

SIZE EFFECT

Most fatigue tests are conducted with carefully controlled specimens (i.e., surface finish, stress raisers, etc.) of relatively small size. As in most areas of engineering, the problem of using such information by extrapolating test results to untested regimes poses a definite concern. In general, with other factors being equal, an increase in section size will tend to cause earlier fatigue failure. This is especially true where notched specimens are used, the method of stress application being irrelevant. Stresses due to bending effects, even for unnotched specimens, also display this same size effect. On the other hand, the use of smooth or unnotched specimens subjected to pure axial loading produces little difference in fatigue behavior as far as specimen size is concerned.

It is certainly reasonable to postulate that the probability of surface defects increases with surface size. As a consequence, the larger the surface area, the greater the tendency for crack initiation. Various numerical factors have been proposed to account for the size effect on fatigue performance but they should be viewed as a first approximation for design purposes and not an exact end to this question. For our purposes, the size correction factor, C_d, will always be taken as unity if $d < 0.3$ in. and 0.85 if $d > 0.3$ in.

STRESS HISTORY OR CUMULATIVE DAMAGE

In most real service applications, the type of controlled stress fluctuations evident in laboratory experiments does not exist. Instead a given stress level may prevail for a certain number of cycles, a different level for another number of cycles, and so forth. Just how to handle such situations has not yet been fully resolved. The earlier attempts to handle this problem are due to Palmgren [4] and, later, Miner [5] and most sources refer to these studies as the Palmgren-Miner cumulative damage theory, or Miner's rule. Where n_1 is the number of cycles involving stress σ_1, n_2 at stress σ_2, and so forth, this rule is usually given as:

$$\frac{n_1}{N_1} + \frac{n_2}{N_2} + \frac{n_3}{N_3} + \cdots \frac{n_i}{N_i} = 1 \qquad (10\text{-}5)$$

or $\sum (n_i/N_i) = 1$ where n_i is the number of cycles subjected to stress σ_i and N_i is the *fatigue life* related to σ_i. This approach has been found to have drawbacks since the order in which the various stress levels are applied does have a decided influence upon fatigue behavior. A more recent approach by Manson et al. [6] appears to overcome the major faults of Miner's rule. To compare these two concepts in detail is beyond the intent of this text and a more thorough discussion may be found elsewhere [7]. In general, if higher stresses are induced during early cycling and are then followed by lower stress levels for a number of cycles, the total cumulative damage is greater than if the stress-cycle combinations were reversed. This may be rationalized by considering that crack propagation would begin more quickly at higher stress levels. Then because of the longer crack length, the influence of the succeeding lower stress levels would be greater than would prevail if the lower stresses had been applied initially when the crack length was smaller.

SURFACE CONDITION

Surface finish, as mentioned in the subsection above, may be viewed as influencing stress concentration on a micro scale. However, the various factors that affect *surface condition* are more closely related to the inducement of surface residual stresses. Since surface tensile stresses have a deleterious influence upon fatigue life, the presence of residual compressive stresses at the surface should be beneficial. Toward this end, certain surface treatments are effective. Mechanical operations such as shot peening and cold rolling with light reductions usually produce residual stresses at the surface. Thermal treatments like flame and induction hardening also tend to produce surface residual stresses in compression and in these two treatments the chemical composition of the material is not altered. Finally, thermal treatments that include a change in the chemical composition, for example nitriding and carburizing, may also produce favorable residual surface stresses. Note that whether the surface condition is altered by mechanical, thermal, or thermal-chemical means, an increase in surface hardness and strength usually results. This may also improve fatigue life and the accompanying compressive residual stress effect can be viewed as a bonus. Finally, the exposure to corrosive environments and the use of chromium plating tend to be detrimental to fatigue life.*

10-3 MACROSCOPIC DESIGN

In the widely used book by Juvinall [3], the subject of fatigue, especially in regard to detail in design, encompasses nearly two hundred pages. To continue with the central philosophy of this text, no attempt will be made here to cover this topic in such depth. Subjects such as statistical aspects, sometimes called reliability, torsional fatigue, fatigue due to combined stresses and fatigue of brittle materials are completely omitted here. In addition, all of the various methods used to handle stress concentration affects are not covered and no attempt is made to introduce the behavior of numerous materials or different microstructures of a particular material in regard to fatigue behavior. Finally, the potential influence of different testing procedures or methods of loading are not covered in any detail; instead, the entire discussion that follows is restricted to the use of one standard approach. If the reader wishes to pursue any of these numerous topics that are omitted, reference to other texts [1, 3, 7] should provide a satisfactory recourse.

SPECIFIC LIMITATIONS

For the remainder of this discussion, approaches to the design problem and conclusions to be drawn are limited to the following restricted conditions:

1. The type of experimental results involved are those obtained from a rotating-beam fatigue machine. Specifically, the use of the R. R. Moore type of machine is of concern. With this device, a carefully produced specimen, of particular physical dimensions and surface finish, is subjected to a pure bending moment. Upon rotation, each surface element experiences a fully reversed condition of tensile and compressive stress per cycle. The sketch in Fig. 10-1(a) displays this situation. Since there appears to be more data available from this particular test method, as compared to any other, it is logical to consider this as the single most important experimental technique if imposed limitations are to be invoked.

2. The behavior of only ferrous materials, and particularly steels, is considered for this discussion.

3. Low-cycle fatigue, where the number of cycles to failure is less than a thousand, is not discussed. It must be noted that in certain

*Plating with zinc and tin appears beneficial.

cases this is an extremely important subject but, for the majority of situations, it is the longer life cases that are of principal concern.

ENDURANCE LIMIT AND THE *S-N* CURVE

By subjecting a number of specimens to different stress levels, where the maximum stress due to the bending effect is simply Mc/I, and counting the number of cycles, N, to failure, the plot of stress versus cycles provides a so-called *S-N* curve. It is almost universal practice to plot these results on log-log coordinates; for steels, a pronounced knee in the curve is generally observed and this is most evident when log coordinates are used. Figure 10-3 typifies this behavior and illustrates the effect of stress concentration, as discussed in Sec. 10-2.

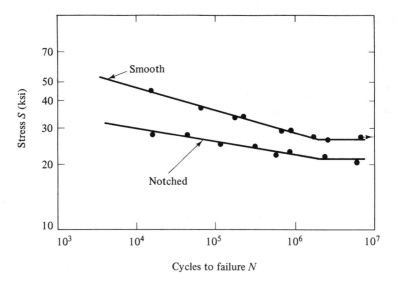

Cycles to failure N

Figure 10-3 *S-N* curves showing a pronounced endurance limit and the effect of notched versus smooth test specimens.

It is essential to realize that there is a fairly wide scatter of the tests results, which is the main reason why statistical aspects enter into the picture. But a crucial point to note is that the scatter of N that results from a number of tests repeated for a given level of *stress* is much greater than the scatter of stress connected with a particular number of *cycles* to failure.

For those materials displaying the behavior shown in Fig. 10-3, any stress level below the horizontal part of the plot would, theoretically, display infinite life since failure would never occur by fatigue. When this behavior is observed, the stress at the knee is termed the *endurance limit* or *fatigue limit*; in this text it will be designated as S_{eb} where the subscript implies this value was found from rotating-beam experiments. The reason for this qualification arises because other types of test procedures will not necessarily lead to the same value of the endurance limit even when all other factors are identical.

Many materials do not display a true endurance limit as indicated schematically in Fig. 10-4. For those situations it is essential to define a fatigue strength, S_f, which *always* corresponds to a certain number of cycles to failure. In fact, the use of S_f, in the same context, must be introduced with materials showing a true endurance limit if the number of cycles to failure in a given design is less than those corresponding to the knee.

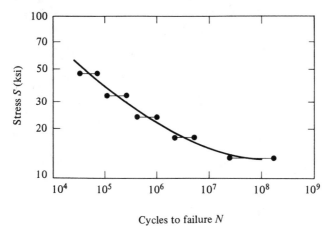

Cycles to failure N

Figure 10-4 *S-N* curve for a material displaying no definite endurance limit.

Fatigue Behavior and Tensile Strength

Figure 10-5 shows the results found when a variety of steels were subjected to fatigue tests. Figure 10-6 compares the endurance limit with the *tensile* strength of a number of steels and cast iron. From such observations, it is now common practice to construct a conservative *S-N* curve using only the tensile strength of the material in those situations where actual fatigue test results are not available. It is *always* better to employ actual fatigue results when they are on hand; only in their absence should the estimation of an *S-N* curve be made using the general relations that follow.

Although there is an obvious scatter in Figs. 10-5 and 10-6, this is not due to any great variation in the tensile strength of a particular steel having a certain microstructure, yet certain trends are apparent. As an arbitrary but reasonable average, the following empirical relations have found use:

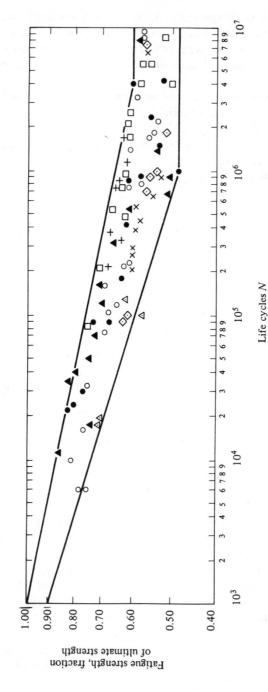

Figure 10-5 Composite of many fatigue test results using rotating beam specimens of steel (reproduced with the kind permission of C. Lipson and R. C. Juvinall (eds.) from "Application of Stress Analysis to Design and Metallurgy," The University of Michigan Summer Conference, Ann Arbor MI., 1961).

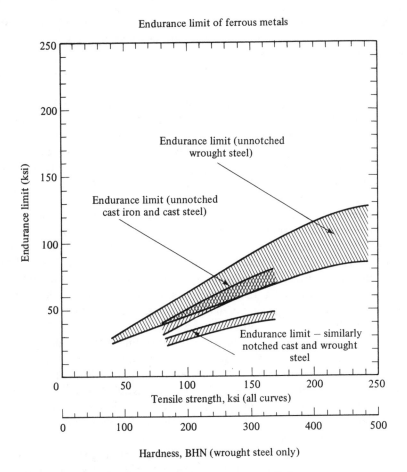

Figure 10-6 Relationship between tensile strength and endurance limit for wrought and cast steels and cast iron (reproduced with the kind permission of C. Lipson and R. C. Juvinall (eds) from "Application of Stress Analysis to Design and Metallurgy," The University of Michigan Summer Conference, Ann Arbor, MI, 1961).

$$S_{eb} = 0.5S_u \qquad \text{where } S_u \leq 200 \text{ ksi}$$

$$S_{eb} = 100 \text{ ksi} \qquad \text{where } S_u > 200 \text{ ksi} \qquad (10\text{-}6)*$$

It must be appreciated that the above relations are used as reasonable estimates in the absence of more meaningful data, yet they have assisted design engineers for many years. The correlations may be revised somewhat by different individuals but the concept still remains. Note also that for *steels*,

*Shigley [7] proposes $S_{eb} = 0.4 S_u$ for cast irons.

in the absence of an accurate value for the tensile strength, there is an acceptable empirical relation between Brinell hardness and S_u. This comes from experimental observation and is given as:

$$S_u = 500\,H_B \tag{10-7}$$

where H_B is the Brinell hardness number obtained when a 3000 kg load is applied to a ball penetrator of 10 mm diameter.* The only time the above relationship can be in serious error is when *primary* or as-quenched martensite is the structure subjected to a hardness measurement. Because of the existence of residual stresses that develop both because of phase transformation effects (there must be an increase in volume for austenite to transform to martensite) and nonuniform thermal conditions as a section fully cools to room temperature, large residual stresses can exist in the piece. Thus a high hardness may not correlate with the potential tensile strength; if the piece is stress-relieved by heating to moderate temperature levels for an adequate time period, the residual stresses can be removed with little if any decrease in hardness. Then, the correlation between hardness and S_u, as given in Eq. (10-7), does prevail. As a final point it is noted that this correlation is quite reasonable whether the steel microstructure is composed of various percentages of ferrite, pearlite, martensite, bainite, or tempered martensite.

Returning to Fig. 10-5, the most conservative line that bounds all of the test points may be defined by constructing a straight line connecting the *ratio* of S_f to S_u of 0.9 at 10^3 cycles with a corresponding ratio of 0.5 at 10^6 cycles. In essence, the estimated value of S_{eb} is taken as $0.5S_u$. Even though this approach is conservative in most instances, individual exceptions can occur in the sense that a particular test point would fall below the sloping portion of the estimated *S-N* curve.

Once a value of S_{eb} is obtained from either actual fatigue test results or, from necessity, by employing the estimation technique discussed above, it is necessary to modify the endurance limit because of the potential influence of the many factors that influence fatigue life; these were discussed in a descriptive sense in Sec. 10-2. The *modified* limit is designated as S_e.

Each modifying factor, if of importance, is used as a direct multiplier of S_{eb}; all tend to lower this basic value which is based upon ideal test conditions. Every such factor is used as an individual coefficient so the combined effects are taken as a pure product. For some, such as the size effect, an average empirically selected number is used, different numbers reflecting certain size ranges. Other coefficients, such as C_s, which introduces the influence of surface finish, are determined using available graphical information.

---

*When $H_B > 450$ it is common practice to obtain a Rockwell C hardness reading and convert this to an equivalent H_B via standard conversion charts. This avoids possible damage to the Brinell penetrator.

In all reference to such situations in this text, the corrected endurance limit is to be found from:

$$S_e = S_{eb}C_sC_d(1/K_f)$$

✓ *Example 10-1.*

A rotating shaft has cross sections of 1.25 in. and 1.75 in. joined by a fillet of 0.125 in. radius; the different sections are supported at the ends by bearings. The shaft is made of steel whose Brinell hardness is 200 and all surfaces are finish machined. Due to the sizes involved, K_t is 1.67 at the fillet and the notch sensitivity, q, for this situation is 0.83. If the maximum stress experienced in the region of concern is 40 ksi, determine the probable number of cycles to failure.

Solution.

(a) S_u is estimated as 500 H_B or 500 (200) = 100 ksi.
(b) From Fig. 10-2, the influence of surface finish gives $C_s = 0.76$.
(c) Since the diameter is greater than 0.3 in., the size correction factor $C_d = 0.85$ from Sec. 10-2.
(d) Since $q = 0.83$, $K_f = 1 + q(K_t - 1)$ from Eq. (10-4) so $K_f = 1.56$.
(e) The value of S_{eb} is assumed to be $0.5S_u$ or 50 ksi from Eq. (10-6).
Thus, the corrected endurance limit is $S_e = S_{eb}C_sC_d(1/K_f)$ or $S_e = 50(0.76)(0.85)(1/1.56) = 20.7$ ksi.

Since this is less than the critical applied stress of 40 ksi, a finite life is expected and the fatigue strength, S_f, is taken as 40 ksi. As discussed in Sec. 10-3, a stress of 90 ksi (i.e., $0.9S_u$) at 10^3 cycles and the value of 20.7 ksi (S_e) at 10^6 cycles may be joined by a straight line on log-log coordinates. This line has an equation of the form

$$S = CN^m$$

For this example, $m = \log(90/20.7)/\log(10^3/10^6) = -0.213$. Using the stress-cycles combination for either known point, C is found to be 392; thus for this example,

$$SN^{0.213} = 392$$

For the applied stress of 40 ksi (i.e., S_f),

$$N^{0.213} = \tfrac{392}{40} \text{ or } N = 45{,}000 \text{ cycles}$$

Note that although other possible correction factors might find use and permit greater empirical sophistication to be introduced, the main concept of this approach is presented by this simplified example.

FLUCTUATING STRESSES
AND THE MODIFIED GOODMAN DIAGRAM

To this point, consideration has been given to the situation where fully reversed stresses prevailed. In many practical situations the imposed stresses may be described by the behavior shown in Fig. 10-1(c); there, an alternating stress is superimposed upon a mean value that is not zero. This is usually described as a fluctuating stress and a brief study of this case will complete this section on macroscopic design.

Numerous approaches to fatigue failure diagrams may be found in the literature. For the purposes of this text, what is commonly called the *modified Goodman diagram** will be introduced. If the reader wishes to gain a perspective of other diagrams, other texts [3, 7] may be consulted. An additional constraint here is that discussion will be limited to ductile metals only.

Figure 10-7 shows the complete diagram of interest. The abscissa denotes the mean stress while the ordinate pertains to the stress amplitude. Assuming no Bauschinger effect, the value of the yield strength, Y, is plotted to scale at the three points indicated as C and connected by straight lines as shown. Point A is located on the ordinate and is the scaled value of the endurance limit, S_e, or the fatigue strength if that is more appropriate. Point B on the abscissa defines the magnitude of the tensile strength, S_u and is connected to A as shown. Because numerous experiments indicate that the magnitude of σ_m when compressive has little influence upon fatigue failure, a horizontal line is drawn from A until it intersects the line indicating yielding at point E.

To fully explain this diagram, several specific examples will be discussed. Consider a loading situation such that the ratio of σ_a to σ_m leads to point F as shown; an extrapolation of line OF intersects AD at point G. When a point such as F falls within the region bounded by $CEAGDC$ and the abscissa,

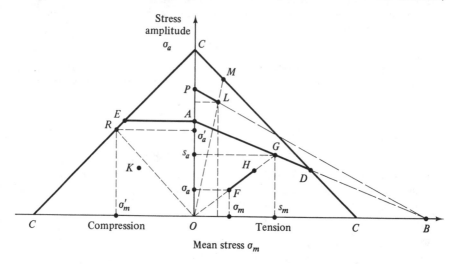

Figure 10-7 Construction of a modified Goodman diagram illustrating numerous situations for the ratio of stress amplitude (σ_a) to mean stress (σ_m).

*See Shigley [7], p. 203, for an explanation.

an infinite fatigue life is predicted. Now line OG defines a locus of points whose ratio of σ_a/σ_m is constant. Suppose a different combination of σ_a/σ_m, that is the same *ratio* but two other absolute values, produced point H. Again infinite life would be expected but greater confidence in such a result must certainly be attributed to the stress conditions defined by F. In this context, a factor of safety is sometimes introduced. The stresses defined by the intersection of OF at G defines stresses S_m and S_a. To define a factor of safety, f.s., related to point F, this is simply S_m/σ_m or $S/_a\sigma_a$ and it is obvious that f.s. must be greater than unity. Since the equivalent values of σ_m and σ_a associated with H must be greater than those associated with F, and since the values of S_m and S_a associated with G are common to both cases, the value of f.s. at point H will be lower than that at F but it will still be greater than unity. In other words, greater confidence, indicated by a higher f.s., would be expected regarding infinite fatigue life for the stress conditions defined by F as compared to H. This should not be surprising since the stress conditions at F are less severe.

If applied stresses produce values of σ_m and σ_a that lead to point K on the compressive side of the ordinate, the explanation just given is still pertinent.

Now consider a situation described by a ratio of σ_a/σ_m that produces point L. Here, there is no meaning to a factor of safety as discussed earlier since such a calculation would give a value less than unity. The physical interpretation for any stress combination that produces a point such as L is that a finite life of a certain number of cycles is expected. By drawing a line from B through L to intersect the ordinate at P, the value of stress defined by P is the necessary fatigue strength, S_f. Following the same procedures used in Example 10-1, where a *finite life* line was constructed, the expected cycles to failure, N, are then found for the stress, S_f. From this concept, any point such as L that falls in the area bounded by $AECDA$ should be handled in this same manner. All such points indicate a fatigue failure of some finite life.

Finally, the lines denoted by $CECDC$ denote yielding in the traditional sense. For any stress state leading to a finding that $\sigma_a + \sigma_m \geq Y$, the point described by the ratio of σ_a/σ_m will fall on or outside the line $CECDC$ and yielding rather than fatigue will dominate. This is true for either the tensile or compressive side of the ordinate. Point R indicates such a situation where corresponding values of σ_a and σ_m are denoted with prime marks to avoid confusion with the earlier use of those terms. Note that the area bounded by DBC is irrelevant in any analysis since any stress state defining a point in that area is outside the line DC, so yielding is expected. Greater detail to this overall approach involving fatigue failure diagrams can be found elsewhere [3, 7]. The handling of specific components such as leaf springs, gears, and bolts plus modifications that arise when brittle materials are involved is covered by Juvinall [3].

Example 10-2.

A round bar of steel having a yield strength, Y, of 40 ksi and tensile strength, S_u, of 60 ksi is to be subjected to a tensile preload of 5000 lb and a fluctuating tensile load that varies from 0 to 10,000 lb. A design analysis indicates K_t is about 2.0 and the best estimate of q is 0.75. If this bar is to have an infinite fatigue life, what should be its smallest diameter, d, if it is ground on the entire surface?

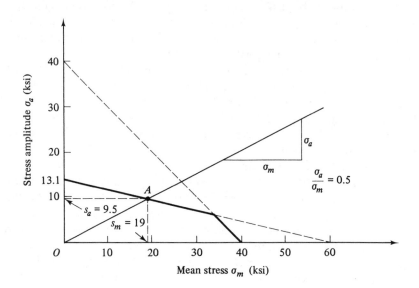

Solution.

(a) From Fig. 10-2, $C_s = 0.9$.

(b) As a conservative value, $C_d = 0.85$.

(c) $K_f = 1 + q(K_t - 1) = 1 + 0.75(1) = 1.75$.

(d) $S_{eb} = 0.5S_u = 30$ ksi.

(e) The estimated endurance limit, $S_e = 30(0.9)(0.85)(1/1.75)$ so with all corrections, $S_e = 13.1$ ksi.

The minimum stress, $\sigma_{min} = 5000/\pi d^2/4 = 6.37/d^2$ ksi.

The stress range, $\sigma_r = 10,000/\pi d^2/4 = 12.74/d^2$ ksi.

The stress amplitude, $\sigma_a = \sigma_r/2 = 6.37/d^2$ ksi.

The mean stress, $\sigma_m = \sigma_a + \sigma_{min} = 12.74/d^2$ ksi, so $\sigma_a/\sigma_m = 0.50$.

Now only the tensile side of a diagram like Fig. 10-7 need be plotted and a line of slope 0.50 drawn from the origin. The intersection point gives stresses of 9.5 and 19.0 ksi respectively so either:

$$\sigma_a \leq 9.5 \quad \text{or} \quad \sigma_m \leq 19.0$$

For $\sigma_a = 6.37/d^2$,

$$6.37/d^2 \leq 9.5 \quad \text{or} \quad d \geq 0.819 \text{ in.}$$

A standard size of $\frac{7}{8}$ in. would be most sensible. Since the choice of d is based upon the *maximum* combination of σ_a and σ_m that predicts infinite life (i.e., point A in the figure), the factor of safety is simply unity.

If a *factor of safety* were specified in the original problem the adjustment would be as follows, using a factor of two for illustrative purposes.

Using the value of $\sigma_m \leq 19.0$, then $\sigma_m \leq 9.5$ since the factor of safety reduces the expected stress level.

Then $\qquad\qquad 12.74/d^2 \leq 9.5 \qquad$ or $\qquad d \geq 1.16$ in.

and a standard stock size of 1.25 in. would be selected.

10-4 FATIGUE AND FRACTURE MECHANICS

In 1860, Wöhler [8] conducted what was probably the first systematic and controlled set of fatigue experiments. For the next century, fatigue studies were concerned with the determination of S-N curves for various solids. The influence of the many factors that modify fatigue strength or the endurance limit and the tying together of all of this information for the purposes of design by the use of plots such as the modified Goodman diagram was studied extensively. Although it was concluded, and properly so, that crack initiation and propagation were the fundamental causes of fatigue failure, fatigue studies were still principally approached empirically.

As the field of fracture mechanics developed, it seemed plausible that crack propagation should be related to the stress intensity factor since this parameter had proved to be of fundamental importance in connection with crack propagation on a macroscopic level. Beginning in the 1960's studies involving the influence of the stress intensity factor on the rate of crack growth were begun and there is now ample evidence to support the general conclusion that the stress intensity factor does govern crack growth, and thus fatigue, in the most fundamental sense.

10-5 CRACK PROPAGATION

Figures 10-8 and 10-9 are schematic illustrations showing the influence of mean stress and starting crack length on the increase in crack length as a function of the number of cycles, N. It is obvious that the crack growth rate, da/dN, increases as N increases and during the latter stages the remaining fatigue life is used up in a very abrupt manner. Since da/dN is related in a functional manner to σ and the crack length a, there should then be a relationship between the rate of crack propagation and the stress intensity factor, K, which itself is related to σ and a. In its most general form, it appears that there is a good correlation to suggest that,

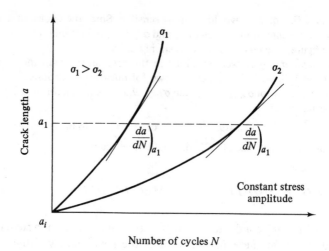

Figure 10-8 Schematic of crack growth versus number of cycles for two 'evels of cyclic stress range but same starting crack length.

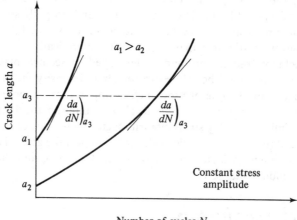

Figure 10-9 Schematic of crack growth versus number of cycles for two different starting crack lengths but same cyclic stress range.

$$\frac{da}{dN} = C(\Delta K)^m \qquad (10\text{-}8)$$

where da/dN is the crack growth rate; C and m material properties; and ΔK = range of stress intensity factor ($K_{\max} - K_{\min}$). Paris [9] suggested that the range of the stress intensity factor was the principal parameter that governs fatigue crack propagation. Figure 10-10 shows one set of results

[10] which, for the range of ΔK versus da/dN involved, the data do fit the behavior expressed by Eq. (10-8). Further support for this concept was provided, as an example, in an interesting study by Swanson et al [11]. They approached the problem by rationalizing that if K was the crucial parameter (i.e., some function of σ and a) then by keeping K *constant* as the crack length *increased*, the crack should propagate at a constant velocity (or constant da/dN). Their findings along with experiments by Paris and Erdogan [12], verified the concept that K was the crucial parameter that controlled crack propagation; in addition, the mean stress or mean K is not as crucial to crack propagation as is ΔK in the region where Eq. (10-8) is applicable.

Before one concludes that the goal of designing for fatigue life is now completely achieved, it must be pointed out that the simple expression given by Eq. (10-8) does not provide a full answer to this problem. The schematic plot in Fig. 10-11 shows the overall behavior of da/dN versus ΔK that is often observed in practice. At low and high values of ΔK, the relationship as given by Eq. (10-8) is not satisfactory and below a threshold value, K_{th}, no propagation occurs. This may be equivalent to stress levels below the endurance limit on an *S-N* curve. In an analogous sense, Fig. 10-11 is similar to creep behavior as shown in Fig. 10-12. There, what is usually taken as the creep rate may define much of the test range in a simple way, but the behavior in the early and late stages departs from the creep behavior displayed during the intermediate range.

An obvious need exists for futher studies to provide the type of information that can be used by designers if fatigue resistance is to be considered from a fracture mechanics point of view rather than the approach described in Sec. 10-3.

To illustrate one attempt to improve the correlation between crack growth rate and ΔK, Forman et al. [13] proposed the following relationship,

$$\frac{da}{dN} = \frac{A(\Delta K)^p}{(1 - R_1)K_c - \Delta K} \tag{10-9}*$$

where A and p are material properties; ΔK equals $K_{max} - K_{min}$; K_c is the fracture toughness; and

$$R_1 = \frac{K_{min}}{K_{max}} \quad \text{or} \quad \frac{\sigma_{min}}{\sigma_{max}} = \text{load ratio}$$

In a physical sense, the denominator in Eq. (10-9) modifies Eq. (10-8) in a manner that makes sense. As R_1 increases with higher load ratios and/or K_c decreases (i.e. lower fracture toughness), the crack rate, da/dN, for a given ΔK, increases. In general this is what is observed in practice, so the drive to

---

*R is the usual symbol in this equation but to avoid confusion with the use of R in Chap. 8, the symbol R_1 is used here.

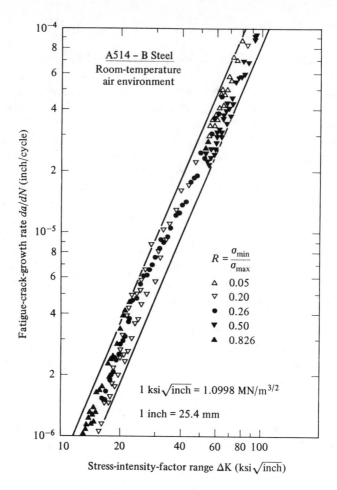

Figure 10-10 Crack growth rate as a function of ΔK for A514-B steel [10].

implement these various factors is rational. Consider an extreme case, where the denominator approaches zero. Then

$$K_c(1 - R_1) \approx \Delta K = K_{max} - K_{min} \tag{10-10}$$

Suppose that $K_{min} \rightarrow 0$, then $R_1 \rightarrow 0$ and $K_{max} \rightarrow K_c$ so catastrophic failure becomes imminent which makes sense. To date, the best correlation between experimental results and Eq. (10-9) occurs when ΔK and the *mean* value of K are reasonably high at the same time. The critical problem seems to arise in deciding just what value of K_c is most appropriate, since it is itself a function of testpiece geometry and crack length; thus Eq. (10-9) should be viewed with some caution.

As a final comment, it should be noted that, in general, materials of higher *strength* possess lower fracture toughness. Thus, the type of reliance upon the

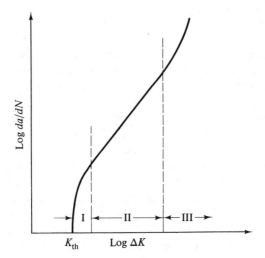

Figure 10-11 Schematic of crack growth rate as a function of ΔK for three regions of ΔK.

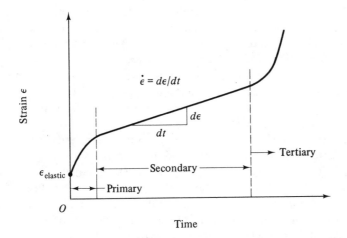

Figure 10-12 Schematic of a typical creep curve.

tensile strength, S_u, being the principal parameter in fatigue design seems questionable *even though* it is still used in this context. It is plausible that a parameter such as (K_c/Y) might be more meaningful.* There is an obvious need for further research in this area.

*See Eqs. (8-58) and (8-62).

Once a component fails, not much can be done about that particular item, yet a careful visual study of the fractured surface may provide useful information. In the broadest sense, the visual inspection and conclusions drawn are combined into what is called failure analysis. Assuming the final conclusions are based upon sound technical grounds, particular modifications in part design or material may be necessitated if future parts are to perform more satisfactorily when exposed to the same service as the fractured part.

On a macroscopic scale, Fig. 10-13 shows a classical fatigue failure. The curved lines are usually called beach marks or clam shell marks; when they are as obvious as those shown, the origin of the fatigue crack can be determined quite definitely. Note too that the appearance of the last portion of the surface to fracture is decidedly different from that portion of the surface resulting from crack propagation. Often, because of oxidation or mechanical rubbing between the two surfaces produced by crack propagation, the patterns shown in Fig. 10-13 may not be so obvious, yet the failure was caused by fatigue.

With the aid of a device such as the scanning electron microscope (SEM) or transmission electron microscope (TEM), much finer details of a fracture surface may be studied. At the higher magnifications associated with these instruments, many microscopic striations (lines) are readily observed *between* two clam shell markings which, themselves, are macroscopic in nature. Thousands and even tens of thousands of striations have been found to exist within the region bounded by two clam shell marks. The striations themselves are indicative of the crack growth *per cycle* whereas the clam shell marks typify a *period* of change in crack growth rate.

Figure 10-14 shows how striations appear at high magnification. Research relating crack growth rate in terms of striation measurements holds much promise for interpreting fatigue upon the most fundamental grounds. It is again emphasized that if striations of the type shown in Fig. 10-14 are observed, the failure is due to fatigue but if such factors as oxidation and surface smoothing due to mechanical rubbing obliterate striations, it cannot be concluded that failure was not *necessarily* due to fatigue. One of the difficulties that also arises is related to striations that are not caused by fatigue in the usual sense. On occasion, damage due to past fracture may cause apparent striations yet the overall failure mode was not true fatigue. Fracture in glass has also caused apparent striations which sometimes intersect at different angles; this is also not true fatigue. In fact, it appears that fatigue striations *never* intersect each other.

Although a thorough coverage of failure analysis is beyond the scope of this book, a few closing words seem appropriate in connection with brittle

Figure 10-13 Macroscopic view of a fatigue fracture surface of a steel axle showing clamshell marking and region of final fracture (axle supplied by D. K. Felbeck, photograph through the courtesy of W. H. Durrant).

Figure 10-14 SEM photograph of striation markings in 1100-0 aluminum (courtesy of W. H. Durrant). Arrow indicates direction of crack propagation. Note increase in crack growth rate (striation spacing) as crack length increases.

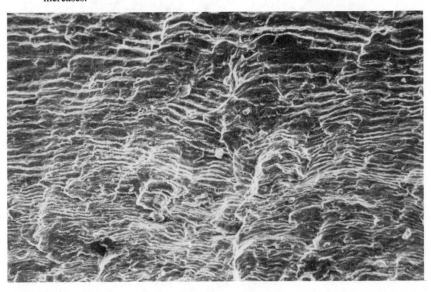

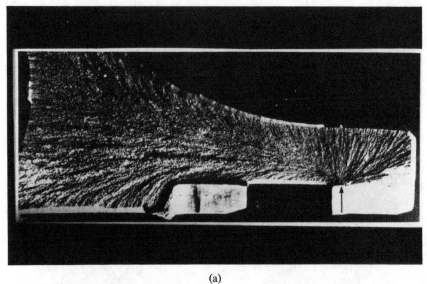

(a)

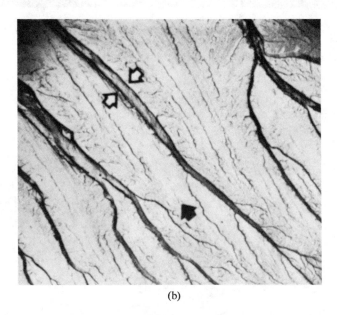

(b)

Figure 10-15 (a) Fracture surface exhibiting a chevron pattern in AMS 6434 steel where fracture initiated at location indicated by the arrow. (b) River patterns on the surface of a brittle transgranular fracture in an Alnico alloy specimen. White arrows indicate a cleavage step while the black arrow shows direction of crack propagation. Both photographs with permission of ASM (from *Metals Handbook*, Volume 10, copyright 1975, American Society for Metals).

failure. Two of the most telling signs are chevron markings and river patterns or markings; these are shown in Fig. 10-15. Not only do they indicate a brittle fracture, but the patterns themselves can be used to identify the origin of the crack. The chevron markings point back to the origin of crack initiation while crack growth is in the opposite direction. River patterns have been so named since they typify a river flow with tributaries joining the major path; crack propagation is in the direction of main flow. The schematics in Fig. 10-16 will assist in this interpretation. Perhaps the most complete coverage of failure analysis in a single source is published by the American Society for Metals [14]. At the very least, it provides an excellent introduction to this important topic.

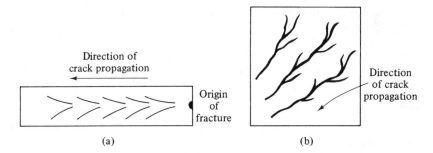

Figure 10-16 Schematics showing chevron pattern (a) and river patterns (b) indicating direction of crack propagation.

As a final comment, the fatigue behavior of polymeric solids is becoming of greater interest now that the use of such materials is finding greater application in load carrying (as distinguished from decorative) designs. Hertzberg [1] provides an excellent summary of this topic along with a number of references.

Example 10-3.

Consider a plate that is 0.5 m wide and one meter in height made of the material described in Fig. 10-10. After being subjected to an alternating stress for a certain period of time, an edge crack of 3 cm length, normal to the plate height and the direction of the applied stress, is observed. The plate was then deliberately fractured and a fracture analysis performed; it was concluded that the 3 cm crack had propagated from a surface flaw. High magnification measurements indicated that the striation width at a distance of 1.5 cm from the crack origin was on the order of 10^{-5} in./cycle. What was the probable magnitude of the applied stress range?

Solution.

From Fig. 8-13(a), $a/b = 1.5/50 = 0.03$ and $h/b = 1.0$. So, $K_I = 1.14 \, K_o = 1.14\sigma\sqrt{\pi a}$. From Fig. 10-10, where $da/dN = 10^{-5}$ in./cycle, the range of ΔK, using the two lines that bound all test data is from 28 to 37 ksi$\cdot\sqrt{\text{in}}$.

Using $1.14 \, \Delta\sigma\sqrt{\pi a} = \Delta K$, where $a = 1.5$ cm $= 0.59$ in.,

$$\Delta\sigma = 28/1.14\sqrt{0.59\pi} = 18 \text{ ksi}$$

or

$$\Delta\sigma = 37/1.14\sqrt{0.59\pi} = 23.8 \text{ ksi}$$

Example 10-4.

A plate whose fracture toughness, K_c, is 60,000 psi$\cdot\sqrt{\text{in}}$. is subjected to a repeated stress cycle from zero to 20,000 psi [see Fig. 10-1(b)]. A small initial edge crack propagated toward the center of the plate as time passed. Because of the plate dimensions, the stress intensity factor, K_I, is approximated by $\sigma\sqrt{\pi a}$.

With regard to Eq. (10-9) the material properties A and p are $5(10)^{-14}$ and 3 respectively.

(a) What crack length would cause catastrophic failure?
(b) When the crack length was 0.30 in., determine the probable crack growth rate.

Solution.

(a) The critical crack length occurs when the value of K_I reaches the fracture toughness, K_c so,

$$\text{Critical } a = \frac{1}{\pi}\left(\frac{K_c}{\sigma}\right)^2 = \frac{1}{\pi}\left(\frac{60,000}{20,000}\right)^2 = 2.86 \text{ in.}$$

(b)

$$\frac{da}{dN} = \frac{A(\Delta K)^p}{(1 - R_1)\,K_c - \Delta K}$$

and since $\sigma_{\min}$ is zero, $R_1 = 0$ and $\Delta K = K_{\max}$, $\Delta K = 20,000\sqrt{0.3\pi} = 19,420$ psi$\cdot\sqrt{\text{in}}$. when $a = 0.30$ in.

$$\frac{da}{dN} = \frac{5(10)^{-14}(19,420)^3}{60,000 - 19,420}$$

$$\frac{da}{dN} = 9.02(10^{-6}) \text{ in./cycle}$$

One note of caution must be added. The magnitude of the constant A will vary if stress units are in ksi rather than psi and care must be exercised in this regard. If all stress units were in ksi above, then A is $5(10)^{-8}$ for the same crack growth rate.

REFERENCES

[1] R. W. HERTZBERG, *Deformation and Fracture Mechanics of Engineering Materials* (New York: John Wiley & Sons, Inc. 1976), pp. 459–61.

[2] R. E. PETERSON, *Stress Concentration Factors* (New York: John Wiley and Sons, Inc., 1974).

[3] R. C. JUVINALL, *Engineering Considerations of Stress, Strain and Strength* (New York: McGraw-Hill Book Company, 1967), pp. 247–49, 262–66, 297–314.

[4] A. PALMGREN, "Die Lebensdauer von Kugellargern," *Z. Verein Deutscher Ingenieure*, **68** (1924), pp. 339–41.

[5] M. A. MINER, "Cumulative Damage in Fatigue," *Trans. ASME, Jour. Appl. Mech.*, **67** (1945), pp. A159–64.

[6] S. S. MANSON, A. J. NACHTIGALL, C. R. ENSIGN, and J. C. FRECHE, "Further Investigation of a Relation for Cumulative Fatigue Damage in Bending," *Trans. ASME, Jour. Eng. Ind.*, B, **87**, 1 (1965), pp. 25–35.

[7] J. E. SHIGLEY, *Mechanical Engineering Design*, 3rd ed (New York: McGraw-Hill Book Company, 1977), pp. 182, 187–88, 201–6.

[8] A. WÖHLER, *Zeitschrift für Bauwesen*, **10** (1860).

[9] P. C. PARIS, "Fatigue—An Interdisciplinary Approach," *Proc. 10th Sagamore Conference* (Syracuse, NY,: Syracuse University Press, 1964), pp. 107–32.

[10] S. T. ROLFE and J. M. BARSOM, *Fracture and Fatigue Control in Structures* (Englewood Cliffs, NJ: Prentice-Hall Inc., 1977), p. 237.

[11] S. R. SWANSON, F. CICCI, and W. HOPPE, *Fatigue Crack Propagation* (Philadelphia, PA: ASTM, 1967), ASTM STP 415, pp. 312–62.

[12] P. C. PARIS and F. ERDOGAN, "A Critical Analysis of Crack Propagation Laws," *Trans. ASME, Jour. Bas. Engr.*, D, **85** (1963), pp. 528–34.

[13] R. G. FORMAN, V. E. KEARNEY, and R. M. ENGLE, "Numerical Analysis of Crack Propagation in Cyclic-Loaded Structures," *Trans. ASME, Jour. Bas. Engr.*, **89** (1967), pp. 459–64.

[14] *Metals Handbook,* 8th Edition, Vol. 10, *Failure Analysis and Prevention,* (Metals Park, Ohio: American Society for Metals, 1975).

PROBLEMS

10-1 A 2-in. diameter bar of annealed 1040 steel is to be used in a situation where loading is fully reversed every cycle of rotation and stress concentration effects are negligible. If the Brinell hardness is 160, determine the endurance limit S_e if the surface of the bar is:
(a) In a hot-rolled condition.
(b) Fully ground.

10-2 Suppose in addition to the conditions in Prob. 10-1, two conditions of stress concentration were to be also considered. These indicate the following:
(a) $K_t = 1.5$ and $q = 0.85$.
(b) $K_t = 2.1$ and $q = 0.77$.

Determine the expected values of S_e again for the hot-rolled and ground surface conditions.

10-3 A steel link connector possesses a yield strength of 0.5 GPa, tensile strength of 0.7 GPa, and an endurance limit of 0.3 GPa which includes all effects of necessary correction factors. The link experiences an alternating tensile stress from 0.02 to 0.1 GPa.
   (a) Is an infinite fatigue life expected? Explain.
   (b) If infinite life is predicted, determine the factor of safety.

10-4 Relative numerical properties of a particular metal part are $Y = 4$, $S_u = 5.5$ and $S_e = 1.25$. Construct the complete modified Goodman diagram; then analyze each of the individual and independent stress situations given below. For each case you are to determine whether yielding, infinite fatigue life or finite life is expected. If infinite life is the result, determine the factor of safety. If finite fatigue life is expected, determine the expected life in cycles to failure, N.
   (a) $\sigma_{max} = 2$, $\sigma_{min} = 1$
   (b) $\sigma_{max} = 3$, $\sigma_{min} = 2$
   (c) $\sigma_{max} = 3$, $\sigma_{min} = -2$
   (d) $\sigma_m = 4$, $\sigma_a = 1$
   (e) $\sigma_m = 2$, $\sigma_a = 3$
   (f) $\sigma_m = 3.75$, $\sigma_a = 0.25$
   (g) $\sigma_{max} = 0$, $\sigma_{min} = -2$
   (h) $\sigma_{max} = -2$, $\sigma_{min} = -3$
   (i) $\sigma_m = -1$, $\sigma_a = 1.25$
   (j) $\sigma_m = -3$, $\sigma_a = 2$
   (k) $\sigma_m = -3$, $\sigma_a = 1$
   For each condition, plot the point defining the combination of σ_m and σ_a on the diagram and label the point with the appropriate letter above. Note that for cases where σ_{max} and σ_{min} are given, σ_{max} is to be taken as the largest algebraic stress.

10-5 A plate made of a pearlitic steel is 6 in. wide, 4 in. high, and contains an initial crack of 0.3 in. centrally located on the 6-in. dimension. An alternating tensile load, perpendicular to the crack, induces extreme stresses of 40 and 20 ksi respectively. This material has a yield strength of 80 ksi, a value of K_{Ic} equal to 135 ksi$\cdot\sqrt{in.}$ and displays a crack growth rate that may be expressed by $da/dN = 3 \times 10^{-10} (\Delta K)^3$, where the growth rate is in./cycle and ΔK has units of ksi$\cdot\sqrt{in.}$
   (a) Determine the crack length that leads to catastrophic failure.
   (b) Determine the average crack growth rate when the crack length increased from 0.4 to 0.5 in.
   (c) Repeat (b) when the crack growth increased from 0.7 to 0.8 in.
   (d) Based upon (b) and (c), discuss how a plot of crack size versus total number of elapsed cycles could be determined.

10-6 A steel pipe of 300 mm outer diameter and 270 mm inner diameter is subjected to internal pressures that vary between 8000 and 3000 psi respectively.

A radial crack of 1 mm length exists at the surface before any pressure is induced. This material has a yield strength of 120 ksi, a fracture toughness, K_{Ic}, of 100 ksi$\cdot\sqrt{in.}$ and a crack growth rate in in./cycle that may be equated to $0.7 \times 10^{-8} (\Delta K)^{2.5}$ where ΔK has units of ksi$\cdot\sqrt{in.}$.

(a) Determine the crack growth rate when the crack is 2 mm long.

(b) At what crack length would failure be predicted?

10-7 The material related to Fig. 10-10 displays a crack propagation relation $da/dN = 0.66 \times 10^{-8} (\Delta K)^{2.25}$, a fracture toughness, K_{Ic} of 150 ksi$\cdot\sqrt{in.}$ and a yield strength, Y, of 100 ksi. A plate of 200 mm width and 400 mm height is subjected to an alternating stress at the ends of the plate and oriented parallel to the height. Due to a flaw at the edge of the plate and located at about the mid-height, a crack propagates at right angles to the alternating load. When the crack is 40 mm long, the plate is removed from service, deliberately fractured and the surface examined. At a distance equivalent to a crack length of 15 mm the average striation width is on the order of 10^{-4} mm while at a crack length of 25 mm the average striation width is 10^{-3} mm. Based upon this information, determine the apparent range of the cyclic stress that prevailed at these two locations.

10-8 Crack propagation data for a given material are reduced to provide equations of the form:

$$\frac{da}{dN} = 3 \times 10^{-18}(\Delta K)^3 \ \frac{in.}{cycle} \qquad \text{as in Eq. (10-8)}$$

and

$$\frac{da}{dN} = \frac{A(\Delta K)^3}{K_c(1 - R_1) - \Delta K} \qquad \text{as in Eq. (10-9)}$$

Note that the exponent for ΔK is 3 in both cases. For the particular conditions σ_{max} is 25 ksi, σ_{min} is 5 ksi and crack length, a, is 0.08 in. the growth rate predicted from either equation is the same. The fracture toughness, K_{Ic} is known to be 50 ksi$\cdot\sqrt{in.}$ and for the geometry involved, $K_I = \sigma\sqrt{\pi a}$. For either equation, the growth rate in in./cycle requires a ΔK having units of psi$\cdot\sqrt{in.}$. From this information:

(a) Determine the crack growth rate predicted from each equation when the crack length is 0.5 in. and the cyclic stress varies from 5 to 25 ksi.

(b) Repeat part (a) if the cyclic stress varied from 15 to 35 ksi.

INDEX

B

Bauschinger effect, 60, 178
Bulge test, 114

C

Composites, 252
  continuous fibers, 254
  definition, 253
  discontinuous fibers, 258
    average fiber stress, 262
    effect of fiber length, 260, 262
    shear stress transfer, 259
  elastic modulus, 256
  fracture toughness concepts, 265

Composites *(cont.)*
  catastrophic failure, 265
  Cook-Gordon debonding, 266
  intermittent shear strength, 267
  Outwater-Murphy debonding, 267
  modified rule of mixtures (MROM), 257
  rule of mixtures (ROM), 254
  strength, 255-262
  volume fractions, 255
Creep, 125 *(see also* Viscoelasticity)

D

Direct compression, 111

Direct compression *(cont.)*
  Cook and Larke method, 111
  Watts and Ford method, 112
Dislocation Theory, 143
  average glide distance, 161
  Burgers vector, 151
  climb, 175
  cross slip, 176
  density, 161
  easy glide, 148
  edge models, 147-151, 157
  forces, 164
    between edge, 176
    between screw, 166
    due to external stress, 164
  Frank-Read source, 172
  maximum theoretical shear strength,
    144
  microcrack formation, 182
  mixed, 147, 148, 150
  movement, 152
  Peach-Koehler equation, 166
  polygonization, 179
  related to macroscopic behavior, 177
  screw models, 147-149, 155
  shear strain, 160
  slip, 144
  strains, 152
    due to edge, 159
    due to screw, 156
  strain aging, 183
  strain energy, 162
  strengthening mechanisms, 177
    cold working, 177
    fine particles, 180
    grain size, 177
    recovery, 179
  stresses, 153
    due to edge, 159
    due to screw, 156
  summary of stress and strain relation,
    186
  yield point, pronounced, 183

Ductility, 99
  per cent elongation, 99
  per cent reduction of area, 99
  prior cold worked metals, 117

E

Elasticity, 47
  constitutive relations, 49
  dilatation, 50
  elastic constants, 50
    bulk modulus, 50, 51
    elastic modulus, 49, 50, 97, 106
    Poisson's ratio, 49, 50, 60
    shear modulus, 50, 52
  homogeneity, 47
  Hooke's law, 49
    one dimensional, 49
    three dimensional, 49, 153
  isotropy, 47
  mean normal stress, 50 *(see also*
    Plasticity)
  Mohr's circle, stress and strain
    relation, 52
  perfectly elastic solid, 48
  plane strain, 56 *(see also* Linear
    Elastic Fracture Mechanics)
  plane stress 56 *(see also* Linear
    Elastic Fracture Mechanics)
  strain energy, 55

F

Failure analysis, 294
  brittle fracture, 294
    chevron markings, 297
    river patterns, 297
  fatigue failure, 294
    beach or clam shell marks, 294
    striations, 294
Fatigue, 272
  description, 272

Fatigue *(cont.)*
  factors of influence, 273
    fatigue strength reduction factor,
      276
    notch sensitivity, 276
    size effect, 277
    stress concentration, 276
    stress fluctuation, 273
    stress history, 277
    surface condition, 278
    surface finish, 275
  fracture mechanics approach, 289
    crack propagation, 289
    growth rate, 289
    range of stress intensity factor, 290
    threshold value of stress intensity
      factor, 291
  macroscopic design approach, 279
    effect of tensile strength, 281
    endurance limit, 280
    fatigue strength, 281
    modified endurance limit, 284
    modified Goodman diagram, 285
    S-N curve, 280
Fracture, 193
  brittle, 194
  Charpy test, 209
  crack stability, 233
  crack velocity, initial, 238
  design considerations, 207
  ductile, 194
  factors that promote brittle fracture,
    195
  Griffith criterion, 201
  maximum theoretical cohesive
    strength, 197
  strain energy release rate, 205
    calibration bar technique, 207
    critical value, 205
  surface energy, 200, 202
  types, 194
    plastic work, 204
Fracture toughness, 224 *(see also*

Fracture toughness *(cont.)*
    Linear Elastic Fracture Mechanics)
    critical plane strain stress intensity
      factor, 209
    compact tension specimen, 242
    compliance, 207, 225
    Gurney's method, 224-230
    influence of specimen size, 239-243
    stiffness, 225
Frank-Read source, 172 *(see also*
    Dislocation theory)

L

Lamé constant, 153
Linear Elastic Fracture Mechanics
    (LEFM), 208 *(see also* Fracture
    toughness)
    critical stress intensity factor, 209
    design aspects, 216
    modes of loading, 209
    plane strain, 209, 220, 239
    plane stress, 210, 220, 239
    plastic zone at crack tip, 212, 239,
      240
    relation between K_{IC} and strain
      energy release rate, 223
    stress equations for Mode I loading,
      210
    stress intensity factor, 209
    values of K_I for various conditions,
      213-219
Lüder bands, 183

N

Necking, 96, 105
    tensile behavior after, 110

P

Plane strain, 56, 209, 220, 239

Plane strain compression test, 115
Plane stress, 56, 210, 220, 240
Plasticity, 58
  compared to elasticity, 59
  deviatoric stress, 66
  distortion energy, 74
  effective strain, 78
  effective stress, 70
  flow rules, 76
  incompressibility or volume con-
    stancy, 77
  invariants, 67
  macroscopic models, 60
  mean normal stress, 60, 65
  Mohr's circles of stress and incremen-
    tal strains, 87
  normality, 81
  octahedral shear stress, 75
  pi plane, 66, 84
  plastic potential, 78
  plastic work, 86
  yield criteria, 66
    Tresca, 67
    von Mises, 69
  yield locus, 63, 72, 73
  yield surface, 65, 84
Poisson's ratio, 49, 60

                        R

Rate effects, 125 (*see also*
    Viscoelasticity)
Recovery, 125 (*see also* Viscoelasticity)
Recovery during annealing, 179
Relaxation, 125 (see also
    Viscoelasticity)

                        S

Strain, 32
  at a point, 33
  at tensile fracture, 117

Strain *(cont.)*
  biaxial equations, 38, 39
  compatibility, 36
  definition, 33
  effective, 78 (*see also* Plasticity)
  engineering or nominal, 42, 97
  hardening, 104 (*see also* Work
    hardening by uniaxial tension)
  incremental, plastic, 44, 77
  instability in tension, 108
  Mohr's circle, 40
  normal, 33
  plane, 56 (*see also* Plane strain)
  rate, effect of, 125 (*see also*
    Viscoelasticity)
  shear, 34
  small in three and two dimensions,
    33, 36
  tensor, 35
  transformation equations, 38
  true or logarithmic, 43, 101
    additive property, 43, 102
    at tensile instability, 108
    relation with engineering, 103
    relation with per cent cold work,
      102
  viscous, 127
Stress, 1
  at a point, 2
  balanced biaxial, 114 (*see also* Bulge
    test)
  biaxial or plane, 15-18
  biaxial space, 72
  cubic equation for principal values, 9
  definition, 1
  deviatoric, 66 (*see also* Plasticity)
  effective, 70 (*see also* Plasticity)
  engineering or nominal, 97
  equilibrium equations, 5, 153
  homogeneous, 3
  invariants, 9 (*see also* Plasticity)
  maximum shear, 18
  mean normal, 50, 60 (*see also*

Stress *(cont.)*
    Elasticity and Plasticity)
    Mohr's circle, 15-27
    normal, 3
    octahedral shear, 75 *(see also* Plasticity)
    principal, 8
    plane, 15 *(see also* Plane stress)
    Orate, effect of, 125 *(see also* Viscoelasticity)
    relaxation, 125 *(see also* Viscoelasticity)
    shear, 3
    sign convention, 3
    space, 83
    tensor, 3
    tensor notation, 3
    three dimensional, 5
    transformation equations, 11
    true or logarithmic, 101
      relation with engineering, 103
    yield, 58

T

Tensile strength, 97
Tensile test, 93
  elastic behavior, 94
  engineering stress versus strain, 96
  instability, 95, 108
  load versus elongation, 93
  necking, 96 *(see also* Neckling)
  per cent cold work, 101
  true stress versus strain, 101

V

Viscoelasticity, 124

Viscolasticity *(cont.)*
  creep, 125, 129, 134, 138, 140
  creep compliance, 129
  four component model, 139
  Maxwell model, 128
  rate effects, 125, 131, 135, 139
  recovery, 125, 129, 134, 138
  relaxation, 125, 130, 135, 138
  relaxation modulus, 131
  relaxation time, 129
  three component model, 137
  viscous model, 127
  Voigt or Kelvin model, 132

W

Work hardening by uniaxial tension, 102
  effect on ductility, 117
  effect on tensile strength, 105
  effect on yield strength, 104
  effective stress and strain, 104 *(see also* Plasticity)
  strain hardening exponent, 104
  strength coefficient, 104
  tensile strength related to work hardening parameters, 108
  work hardening equation, 104-107

Y

Yield strength, 97, 103
  as influenced by work hardening, 104
  measurement, 98
  upper and lower yield points, 99, 107, 183